Table of Contents

Preface

When most people break a bone, they go to the doctor right away to get medical treatment. They listen to their doctor's advice about what they should and shouldn't do to help the healing process along. They get plenty of rest and avoid putting too much stress on the broken bone so it will mend as quickly as possible, and if needed, they go to physical therapy and gradually strengthen the limb over time until they've made a complete recovery. Most importantly, they don't pretend the injury doesn't exist or play it off like it's not a big deal while the problem gets worse. And yet, far too often, we find ourselves failing to treat mental health issues with the same seriousness we would for physical health issues. We avoid telling anyone about the issue, especially a psychologist. We often put off treating the issue at all, and the longer we ignore it, the greater the toll it takes on our well-being.

Your mind is one of the most important tools you have. If your mental health is suffering, you're going to have a hard time staying on top of your physical health too. You might find that it's harder to maintain relationships and to make new ones. You may have a difficult time focusing on your work, too distracted with anxieties and negative thoughts to make your best efforts, potentially jeopardising your job as a result. You'll likely feel less confident and often less happy on a daily basis. You may also have trouble sticking to regular sleeping and eating habits, which could lead to increased weight gain, high levels of stress, and impaired cognitive abilities. In short, your body functions as a connected system, and if any one part of that system falters, you'll feel it everywhere else.

Luckily, you don't have to live with poor mental health forever. You can embrace a more positive mindset and let the stress you always carry with you melt away, leaving you happier and healthier with a new lease on life. In *The (Mental) Health and Fitness Coach*, I'll show you exactly how you can begin making improvements to your mental health and feeling more like your old self again. You'll learn how to identify the barriers that hold you back and how to break through them, using your environment, habits, and positive attitude to your advantage. You'll also discover invaluable methods for easing anxiety and ensuring your happiness in the long term.

As a mum of three boys, all under five years old, I used to frequently find myself stressed out, both in the home and outside of it. I suffered from anxiety, which wasn't helped by my highly stressful management position. I ended up spiralling, stuck in a physical and mental state that only got worse with each added source of stress. I knew I had to regain control of my life, so I started studying behavioural change and the benefits of a positive mental attitude on mental health. As I implemented these principles into my own life, I noticed a gradual change that grew and grew each day I stuck to my new healthy habits. I strengthened my resolve and started forging my own path to success, leaving my anxieties in the past where they belong.

My ethos involves helping people become their ideal selves by addressing both mental and physical health. I want to help you create a life of balance between your nutrition, your physical exercise, and your mental well-being, all of which are core components of a happy, healthy, and satisfying life. There is plenty of information available on these topics, but there are so many different sources and viewpoints that it can be confusing to know what you should listen to and where you should find all the necessary information. Experiencing this confusion myself, I started studying so I could become well-versed in these topics and provide a useful resource with everything you need to know to improve your life. My *Health and Fitness Coach* series seeks to do exactly this, compiling everything I've learned into an integrated plan with steps for improving every aspect of peoples' lives.

I know first-hand just how helpful these simple steps for revitalising your mental health can be, and I want to pass them along to as many people as possible so that you too can become more knowledgeable about the inner workings of your mind and more in control of your thoughts. Through the information within this book and additional resources you can find at https://thehealthandfitnesscoach.co.uk/ or https://thehealthandfitnesscoach.com, you will be empowered to work towards becoming the best version of yourself.

We have to live in our minds 24/7, so we need to treat our mental homes with dignity and care. We can't expect ourselves to succeed if we don't give ourselves the right tools, and that means taking care of our mental homes just like we would change the sheets and water the plants in our physical homes. I will be your mental health fitness coach, guiding you towards the positive changes you'll need to make to achieve this level of mental strength so you can pursue personal and professional success. As you follow along with this book, you will build healthy habits that will enable you to look after your mental well-being, manage your stress and anxiety, and become truly mentally fit.

Ready to get started? Your first training session begins right now.

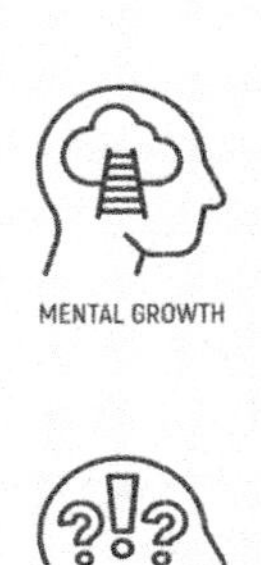
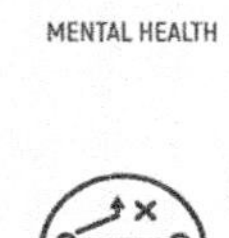

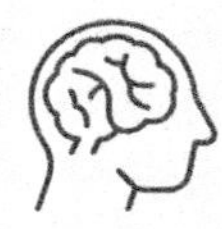
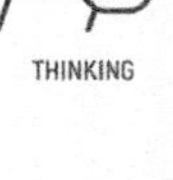
MENTAL GROWTH
MENTAL HEALTH
NEGATIVE THINKING
THINKING

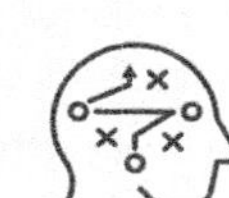
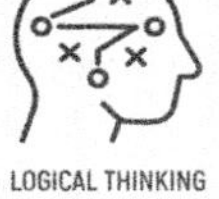

EMOTIONAL REASONING
LOGICAL THINKING
OBSESSION
INNER DIALOGUE

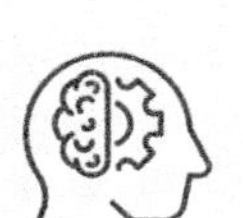

EMOTIONS
THINKING PROCESS
MUSICALITY
BALANCE

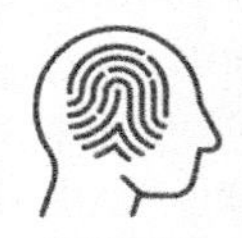
POSITIVE THINKING
BRILLIANT THOUGHT
CREATIVE THINKING
SELF IDENTITY

Chapter 1: Introduction to Mental Health and Fitness

Your brain is one of the most complex organs in your entire body. It controls many different functions in your body, both conscious and unconscious. Whether you're running a marathon or you're fast asleep, your brain is hard at work regulating everything that goes on throughout your body and inside your head. It's no wonder, then, that an interruption in your brain's regular processes can have a significant impact on the rest of your body.

Your brain chemistry is closely tied to your mental health. Different hormones and other chemicals can affect how your brain processes various stimuli, which are anything in your environment that you react to. If you've been struggling with your mental health, you might have noticed that something that once made you happy doesn't have quite the same effect. Maybe you used to enjoy watching your favourite movie or doing a fun activity like going out to a restaurant or playing soccer, but you no longer feel quite so interested in those same things, and they don't bring you the joy they once did. On the other hand, you might be more reactive to your environment than ever before if you've been experiencing anxiety. You may find that stimuli that didn't bother you much before now leave you feeling nervous and often fearful at the slightest provocation.

These changes can be confusing and troubling, as they interfere with your ability to lead a happy, fulfilling life like you once did, and they may seem to have no clear cause. However, the cause is often connected to the different hormones and other systems at play in your brain. As you begin to learn more about how your mind functions and what impact this can have on your mental health, these changes should start to feel less frightening, since you will instead feel empowered to start regaining control over these difficult thoughts and feelings.

First, let's begin by taking a closer look at your brain and how it affects both your mental and physical well-being.

How Your Mind Operates

Your brain is like the command hub for the rest of your body. It is an intricate system of different regions that are responsible for controlling various conscious and unconscious processes. It's so complex that we don't fully understand it yet, and it may take many more decades of research before we do. However, this doesn't mean that everything that goes on in your brain has to be a complete mystery to you. Having a better picture of the general processes that influence your mental health leaves you better equipped to address any issues you may be experiencing, and you don't have to be a neuroscientist to understand the basics.

When you're having an especially bad mental health day, it can be tempting to blame yourself. You might think dark thoughts, such as assuming you're broken and you cannot be fixed. You may even wonder if there's anything wrong with you at all, and if you're just exaggerating or making things up, especially if other people have been dismissive of your feelings before. If you don't know about any of the processes going on in your brain that could be contributing to conditions like anxiety and depression, these kinds of thoughts are all too common. When you start to understand exactly what's happening under the hood when your mood is poor or when your mind is racing, you'll have an easier time recognising the issues that abnormal brain chemistry or neurotransmitter functions can cause. This then makes it easier to accept that, while you might not feel very good now, this doesn't have to be permanent. You can start training your brain to better handle stress and anxiety, and use mental health treatments that really work. Everything starts with developing a better understanding of your brain.

Parts of the Brain

Different areas of your brain are responsible for performing different tasks. For example, the cerebellum, which is located at the base of the brain, is responsible for controlling your balance and coordination. The cerebrum, or the main part of your brain, is made up of different lobes. These include the frontal, parietal, temporal, and occipital lobes, which are different regions that accomplish a variety of related tasks. To keep things simple, I'll focus on the areas of the brain that impact your mood and thought patterns, as these have the strongest ability to affect your mental health. Still, it's important to recognise that just because one area of the brain controls a certain function, that doesn't mean it's acting alone. Your brain is in constant communication with itself and the rest of your body, and these areas can positively or negatively impact each other.

Your frontal lobe is responsible for managing things like emotions, social interaction, problem solving, and impulse control. If you have poor self-control that exacerbates your mental health issues, this may be because your frontal lobe isn't properly regulating your executive functions. You might have difficulty achieving your goals because you frequently find yourself distracted, avoiding work for more enjoyable activities that take your mind off your stress. Poor executive function can manifest as an inability to control your emotions or impulses, so you have trouble focusing on tasks that aren't engaging, and you may have more emotional outbursts when things don't go as planned. This can make it incredibly hard to deal with stressful situations, big changes, and tight deadlines, especially if you're already struggling to stay motivated due to anxious or depressed thoughts.

Another important region of your brain involved in emotional control is the amygdala. Your amygdala assists in regulating emotion, especially emotions with a strong connection to certain memories. For example, if you have good memories of watching a certain TV show with your parents, you might feel happier just by hearing that show's theme song. This can be true for stimuli, or environmental factors, that make you remember negative memories too. If you were in a car accident, just the thought of getting into a car can make you panic because your amygdala facilitates such a strong mental connection between your memories and your emotions.

When you encounter something that upsets you, the logical and emotional sides of your brain have a bit of a tug-of-war. While your frontal lobes in your cerebral cortex try to process the stimulus and react to it rationally, your amygdala is responsible for automatic emotional responses that trigger your "fight or flight" reflex. This can be incredibly helpful if you're a caveman trying to escape or fend off a mountain lion and you need a boost of adrenaline, but it's much less helpful when you encounter an otherwise benign stimulus that makes you stressed or panicked because it reminds you of past experiences. This experience is known as an "amygdala hijack," first coined in Daniel Goleman's book *Emotional Intelligence*. In it, he writes that during an amygdala hijack, "some emotional reactions and emotional responses can be formed without any conscious, cognitive participation" (Goleman, 1995, p. 18). Instead, reactions are almost entirely dictated by emotional memory, which operates without any interference from reason and logic. This is why we can have emotional outbursts and later ask ourselves, "What was I thinking?" or "Why did I do that?" We have our amygdalas to blame. Thinking more rationally about these situations before we react can help us interrupt an amygdala hijack and keep our emotions in check.

Each of the different parts of your brain that affect your emotional and behavioural responses are part of your limbic system. This includes your amygdala and your hippocampus, which stores memories. The limbic system forges the strong connection between memories, emotions, and behaviours that can cause you to act erratically when stress, worry, and fear take over. The connection between these different centers in your brain is made possible by chemical messengers known as neurotransmitters.

Neurotransmitters

Neurotransmitters function as a messaging system for your body. You can think of them as mental mail carriers. They are the trucks that carry 'mail,' or different signals, between neurons and cells. You have many different kinds of neurotransmitters all throughout your body, but they're most commonly found in the brain.

Different chemicals in your brain act as neurotransmitters. A disruption in your regular brain chemistry can interfere with chemical signalling inside your brain, which can, in turn, trigger or exacerbate mental health issues.

Your Brain Chemistry

Various chemical neurotransmitters have been linked to mental health. Having imbalances in these chemicals, which are naturally produced by your brain, could make it harder to fend off the symptoms of conditions like depression and anxiety. Notable chemicals that have been associated with mental health include norepinephrine, dopamine, and serotonin.

When you encounter something that stresses you out, your body produces heightened levels of norepinephrine. This is the neurotransmitter and hormone that is responsible for your "fight or flight" response. It increases your heart rate, blood pressure, and blood sugar, so you have more energy to fight or flee. When norepinephrine is present in short bursts in appropriate amounts, it can be helpful. For example, if you need to run away from something dangerous, your reflexes are heightened and you can run for far longer than you could under normal, non-stressful circumstances. This can save your life in a tense situation. However, we rarely encounter these kinds of situations in our daily lives. Instead, you might experience a lot of stressful situations at work or university, which means your blood pressure remains too high for an extended period of time. In addition to making it harder for you to relax and process situations calmly, this can also put strain on your physical health.

On the other hand, if your norepinephrine levels are too low, this can be a problem too. You may not have the energy you need to get through the day, leaving you feeling foggy, lethargic, and disinterested. Low norepinephrine levels are often associated with conditions like depression and ADHD.

This problem is exacerbated if you also have unusually low serotonin levels. Serotonin assists in regulating your mood and reducing anxious thoughts. Many people with depression have low levels of serotonin, which points to a correlation between these two factors.

Therefore, serotonin is a common target of medications prescribed to help reduce the symptoms of depression, such as selective serotonin reuptake inhibitors (SSRIs) and serotonin-norepinephrine reuptake inhibitors (SNRIs). These medications keep the nerve cells in your brain from reabsorbing serotonin once its job of delivering a message is done, which means there are higher levels of serotonin in your brain as a result. Of course, medication isn't the only way to improve serotonin levels or your mental health as a whole. You can also focus on changing your behaviours, thought patterns, and even environmental factors to ease the symptoms of poor mental health.

Another notable brain chemical that has been connected to mental health is dopamine. Higher levels of dopamine cause you to feel happier, while lower levels can negatively impact your mood and make it harder for you to focus. Dopamine is meant to help us learn and reinforce positive behaviours, making it a natural reward system. Unfortunately, we can also experience dopamine spikes when we do things that are bad for us like eating junk food or playing video games instead of getting our work done, which trains our brains to seek out these experiences over healthier ones. While some medications can increase the amount of dopamine in your brain just like serotonin, you don't need to completely rely on them to regulate your dopamine levels either. Adopting a healthy routine you can feel good about and reducing sources of stress in your life can have a significant impact on your brain chemistry, whether used alone or combined with medications or professional help. Over time, you can shift your mental reward system away from the quick-fix of sugary snacks and addictive behaviours and towards a routine that helps you feel healthier, happier, and more at peace with yourself.

Combining the Mental and the Physical

It's a common misconception that mental and physical health are entirely divorced from one another. In truth, there is much more crossover than you might initially assume. Poor mental health can cause serious harm to your physical well-being, and failing to take care of yourself physically often means failing to take care of yourself mentally. The reverse is also true: practicing good habits for physical health like eating right and exercising can help you boost your mental health, and caring for your mental health will make it easier to look after yourself physically as well.

Stress is one of the biggest factors that complicates your attempts to maintain good mental and physical health. You're probably used to the feeling of stress weighing you down mentally, especially if you have a hectic lifestyle, a demanding job, or an unusual living situation. What you might not realise is that anything that causes you mental stress is also taking a toll on your body. As previously mentioned, high stress levels can raise your blood pressure, which forces your body to work harder each day. High blood pressure caused by stress can increase your risk of developing conditions like heart disease and kidney disease, and it can also leave you more likely to have a heart attack or stroke.

Continually living in a high-stress environment can also make it harder for you to make healthy choices. You might find it more difficult to focus on your work or to exercise because you'd rather be doing something fun that provides a more readily-available boost in your dopamine levels. You may reach for the sugary junk food instead of a meal or stop to buy fast food on the way home, simply because it's easier and you don't have the mental capacity to worry about dinner. You might have less patience for difficult conversations with family and friends because of your stress and low mood, which can damage your relationships.

Luckily, caring for yourself physically can be a powerful tool for reducing your stress levels and improving your mental health. According to the Anxiety and Depression Association of America, "exercise and other physical activity produce endorphins—chemicals in the brain that act as natural painkillers—and also improve the ability to sleep, which in turn reduces stress" (Anxiety and Depression Association of America, n.d., para. 5). This showcases just how connected your physical and mental well-being are and how important it is to take appropriate steps to deal with stress and other sources of mental health issues in your life.

How to Begin Rewiring Your Brain

Beginning to unravel the unhealthy coping mechanisms, toxic life situations, and other detrimental behaviours that contribute to poor mental well-being can be a long and sometimes difficult process. Think of retraining your brain and improving your mental health like the process of recovering after an injury. If you keep pushing yourself the same way you've been doing up to this point, you won't give yourself a chance to heal. You'll most likely only make the injury worse. But if you're kind to yourself, you stick to your therapy whether it's physical or mental, and you seek help when you need it, the injury will mend over time.

There are many different methods you can use to improve your mental health. You can use strategies like improving your communications skills, assessing yourself, reducing stress in your daily life, changing your environment, replacing bad habits with good ones, getting restful sleep each night, and supporting your physical health to achieve a more fulfilling mental state. Not all of these methods will yield immediate results, but they all encourage making positive changes in your life that help you better understand and deal with your mental health problems. As you reflect on yourself and start making improvements, you'll begin to leave that mental rut you've been stuck in so you can enjoy a more positive mindset and all the benefits that follow.

MIND CENTER

CREATIVE POWER

CREATIVE MIND

BRAIN POWER

Chapter 2: How to Understand Yourself

Before you can begin strengthening your mind and improving your mentality, you must first develop a baseline understanding of where you are at right now. Think of this like weighing yourself before you start a new diet or workout program. You need to know where you're starting from so you can decide where you're going and track how much progress you're making towards your goal.

This process begins with developing a deeper understanding of yourself. It's very easy to go through life without really understanding who you are and why you feel the way you do in certain situations. For example, if you get angry when you feel like someone's talking over you, even to the point that it seems irrational to others, you might just accept this as fact and move on. You might even decide that you're completely right and it's everyone else who is being disrespectful and purposefully rude. While it's never nice to feel like you're not being listened to, this doesn't mean that your excessively angry reaction is an appropriate response either. But you cannot work to correct this issue until you know it exists and you have a better idea of why you might react this way. This means reflecting on yourself and trying to learn more about who you are from a non-judgemental angle.

It's also important to try and understand what barriers are holding you back from self-improvement. These may be situational barriers, like a negative environment that reinforces discouraging thoughts and a stressful job that takes up all your time, or internal barriers, such as your own self-doubt and previous experiences that left you feeling afraid or frustrated in certain situations. If you don't confront these barriers in whatever form they appear, they're going to prevent you from growing as a person and improving your mental well-being.

Pit these barriers against the parts of yourself that serve to create a foundation of positivity. Look for opportunities to improve your own skills for greater mental strength. Focus on creating a deep-seated sense of self-love and self-worth that won't crumble when you encounter adversity. Everything starts by looking inward, identifying what needs to be fixed, and replacing these problem areas with genuine positive growth and change.

Identifying Your Barriers to Change

Many people who experience anxiety find that even though they can rationalise a situation in their minds, they can't necessarily control their fear. For example, anxious thoughts often stem from the mistaken belief that everyone is hyper-focused on your every move and harshly judging you for any perceived flaws, even if you don't treat others with this same level of criticism. You might be fully aware that these thoughts are irrational and that in truth, no one but you really remembers when you slip up and say or do the wrong thing, especially if they're not personally acquainted with you. Even knowing that, you might still be unable to get your feelings of anxiety under control because there are other barriers impeding your ability to change.

Knowing the right thing to do to improve your mental health doesn't always mean you're ready and willing to do it. You might avoid something like meditation, even if a therapist tells you it can help you calm erratic thoughts. You might resist the idea that physical changes like improving your dieting and exercising habits could contribute to physical change. Sometimes you might not even avoid these healthy behaviours cognitively but instead come up with justifications for avoiding them so you can still feel like you're doing the right thing. Unfortunately, if you never change anything about your thought patterns or how you treat yourself, your mental health isn't going to change either. In addition to knowing what you should be doing to improve your mental fitness, you'll also need to address the barriers that prevent you from doing so.

As previously mentioned, these barriers can come in many different forms depending on your personal experiences, current situation, and general disposition. Some of these barriers arise from external conflicts and life circumstances that may or may not be within your control. Others are entirely internal, which can be harder to break down but which are always within your power to eliminate. In either case, focus on changing what you have control over—your reactions to tough situations and stressful moments—and keep making small, positive changes to chip away at these barriers bit by bit. The more progress you can make on removing these barriers from your path, the easier it will be for you to make the positive mental, emotional, and physical changes necessary for mental strength and a positive mindset.

External Barriers

Many barriers come from external factors in our lives. These often arise from difficult situations or certain people in our lives, both of which can hold us back from improving. For example, maybe you have a very demanding schedule that subtly encourages you to lean into your worst habits for mental health. You might pull longer hours at work, which leaves you grumpy and exhausted. Your exhaustion makes it harder to be productive at your job and at home, and negative feedback from your boss reinforces the self-doubts you hold about yourself. Your poor disposition means you're more likely to pick fights with friends and family, which can damage your supportive relationships. You might instead choose to isolate yourself to prevent lashing out, which is often bad for mental health. Compound this with other components like relying on cheap, fast, and unhealthy takeaways for meals and foregoing sleep in favour of trying to get more work done, and you can see just how many barriers can crop up from something as seemingly straightforward as a few busy days at work.

The same is true when there are people in our lives who reinforce negativity and lead us to think the worst of ourselves. They might reinforce negative thoughts you have about yourself, or they may fail to understand your struggles with mental health and put a lot of pressure on you to pretend they're not there. If you're serious about improving your mental health, keeping these kinds of people around can make this goal nearly impossible. Not everyone in your life who is a barrier to good mental health says hurtful or insensitive things on purpose, and for those who do it accidentally, you may be able to break down these barriers by having an honest conversation about why their actions and comments have negatively affected you. Others may be more toxic, knowingly making these kinds of comments, and a more permanent separation may be necessary, as we'll discuss later on.

Internal Barriers

Just as barriers can come from external factors such as your environment and different stressful situations, they can come from internal factors too. These factors may be the products of a previous negative experience that is discolouring many of your current interactions. For example, maybe you had a bad breakup recently, or perhaps you were even in a toxic or abusive relationship. This can make the idea of trying to start dating again a terrifying one. If your partner belittled you, harming your sense of self-worth, you may start believing that their words must have been true and you're simply unlovable. If someone ever does show genuine interest and kindness, it can be harder to accept and reciprocate these affections, because you have already subconsciously convinced yourself that you're not worthy of them. This is an incredibly damaging belief to hold about yourself, and you might not even recognise that it's become part of your internal monologue, let alone that it stems from this previous relationship.

These kinds of internal barriers can manifest from past experiences that may seem fairly benign on the surface as well. Many people decide they simply can't have mental health issues because their childhoods were "completely normal," but even minor childhood events can affect our development significantly. Maybe you were a loud child who was frequently shushed by your parents and teachers, and now you have trouble speaking up and regaining your confidence in social situations because your mind is convinced someone will find you annoying or aggravating again. Maybe you experienced poverty and food insecurity for a while, and now you tend to stress about finances to an excessive degree. You might feel less confident about sharing your passions and hobbies if you were discouraged from pursuing them in the past. Emotional barriers such as fear, uncertainty, doubt, and self-hatred can all take their toll on your mental health and keep you from improving if you never address them.

Identifying the internal barriers that hold you back specifically can be more difficult than understanding the effect of external barriers. You may have to do some self-reflection, and in some cases, you may benefit from speaking to a therapist or another mental health professional who can help guide you through confusing thoughts and feelings. It's often difficult to confront trauma, whether that trauma seems minor or major, but you cannot start to heal until you understand what is holding you back.

Barriers That Discourage Seeking Treatment

In addition to the barriers that actively harm your mental health, some barriers are more about limiting your ability and willingness to seek mental health treatment. Even if you know you're struggling, you might feel like you're unable to reach out, or you're uncertain about trying some of the techniques that could help you improve. A 2015 study published in the journal *Professional Psychology: Research and Practice* identified common barriers that included external factors such as "time and effort concerns, logistical concerns, [and] financial concerns," as well as internal factors like "confidentiality concerns, stigma, [and] the ethic of self-reliance," (Becker et al., 2015, p. 504) all of which make it harder to pursue the development of a stronger mental state.

Real change takes time and effort. This is true for change in all its forms. Still, it can make the process of self-improvement feel more daunting if you know it's going to take some work before you see results. While there's little way around this requirement, there are two things to keep in mind that can help reduce its power as a barrier. First, while it takes a while to build up to big changes, you will likely start to notice incremental changes as soon as you start practicing better habits for your mental health. And second, the immense benefits of embracing change are well worth the time and effort you will spend achieving them.

Logistical and financial concerns can make seeking traditional treatment more difficult. It's hard to get to a therapy appointment if you don't have a car or another form of transportation, or if you work a demanding job that limits the free time you have to attend.

Additionally, sessions can be expensive, especially if you are going every week or so. Luckily, there are a few ways to lessen the effect of these barriers as well. You might find that speaking to a mental health professional online or over the phone fits better into your lifestyle, and if money is tight, you can make use of local government programs that help connect you with mental health services in your area. You can also focus on improving your habits and interrupting negative thought patterns with the help of this book and other resources rather than relying on traditional therapy if you can't afford it.

Internal factors, like the stigma surrounding mental health issues, can be especially powerful deterrents for seeking support and implementing helpful strategies. It's tough to talk about mental health with other people, as it requires you to be very vulnerable and people may react in negative ways. The UK-based Mental Health Foundation has pointed out how "media reports often link mental illness with violence, or portray people with mental health problems as dangerous, criminal, evil, or very disabled and unable to live normal, fulfilled lives," (Mental Health Foundation, 2015, para. 12) which can then affect how the average person views someone struggling with mental health issues. These stigmas can make it very difficult to continue pursuing self-improvement, even if you know it's the best choice for you. While public attitudes are changing, remember that you can still work to improve your mental health without having to inform everyone in your life of your struggles if you don't feel comfortable doing so.

Additionally, remind yourself that conditions like anxiety and depression don't make you weak, and they're just as important to treat as any other health condition.

Social stigmas also help perpetuate the "ethic of self-reliance," which suggests that asking for help is something shameful that you should avoid at all costs. Of course, this is far from the truth. No one can make it through life on their own, and it's completely understandable to need help sometimes. Self-reliance can be powerful in the right circumstances, but it should never come at the cost of your health, physical or mental. When pride becomes a barrier to self-improvement, it's better left behind you.

Now that you know the types of barriers that can get in the way of personal growth, consider which ones might apply to you. You may have just one or two barriers, or you may have many that affect you to varying degrees. Once you've identified your own personal roadblocks, you can work on confronting and smashing through them, leaving you free to start changing your thoughts and behaviours for the better.

Breaking Through Your Barriers

Mental health barriers prevent you from pursuing effective treatment strategies. They hold you back from self-improvement and make it harder for you to live a satisfying, fulfilling life free of constant stress and worry. It's critical to start breaking down your barriers once you identify them, as this will enable you to address any struggles you're experiencing and start to heal from them. There are many ways to begin the healing process, but they often start by targeting the barrier of the stigma that surrounds mental health issues.

Overcoming Stigma

Ideas that are frequently repeated in popular media and in discussions throughout our communities become part of our society. They form social mores and taboos, dictating which behaviours and opinions are socially acceptable and which are not. Unfortunately, mental health has long since become a bit of a taboo to speak about, especially for people who want to be perceived as strong. Struggling with mental health is viewed as weak and, in some cases, even dangerous. Media depictions reinforce this and affect the way that society views mental illness as a whole. This can make it very hard to take your mental health issues seriously, let alone to start addressing them, as to admit you have them at all would be to admit weakness.

Of course, in truth, there is nothing about having mental health issues that makes you weak, nor anything that should lead you to ignore them and pretend they don't exist in the first place. This only allows them to grow stronger, increasing the effect they have over your life. If you ignore an issue, it doesn't go away; it just lurks under the surface, striking at the least opportune times. If you don't develop a toolset for dealing with moments of worry, stress, fear, and self-loathing thoughts, you won't be prepared for these feelings when they make themselves known. It is only by acknowledging the problem that you can begin to correct it, which means overcoming the stigma surrounding mental health issues and taking your mental health as seriously as your physical health.

Part of overcoming stigma must happen internally, as you challenge harmful thoughts that reinforce negative stereotypes about mental illness. However, you can reinforce this practice by seeking external support. Spend time with people who understand what it's like to navigate mental health issues, and who can offer you advice without belittling your struggles. Consider joining a support group, online or in person, whose members can empathise with your experiences. Remember that in most cases, mental health stigma arises from a lack of understanding of the issues you're facing, not from factual information. Make an effort to better educate yourself and you will find many of the stigmas you previously held because of societal beliefs falling away.

Addressing Other Barriers

Even once you've addressed mental health stigmas, other barriers may complicate your journey to improve yourself. Take the example of having a busy job, which in turn, leaves you with a lot of stress and little time to practice good techniques for managing your mental health. Here, you will need to practice some problem solving to devise a solution that works for your situation. You might benefit from practicing better work-life balance, which would give you more free time for de-stressing and make work less overwhelming. You may also be able to speak to your boss about your concerns and ask for support, or you can focus on keeping your non-work obligations to a minimum. Any way you can ensure you have the time and energy to practice good mental wellness strategies will increase your chances of making smart, healthy choices whenever possible.

You may also have to spend some time reflecting on the previous experiences and negative thought patterns that contribute to anxiety, depression, and other mental health concerns. Make an effort to understand not only what you're feeling but also where these feelings are coming from. Consider why a certain situation makes you nervous, frustrated, or defensive, and when you might have encountered something similar before that has made you more wary of these circumstances. Once you understand where that fear and resentment is coming from, you can approach the problem from a more logical, level-headed angle. Making an effort to understand and accept yourself even while striving to improve is key for breaking down your barriers to growth.

Practicing Positivity and Self-Love

Wanting to grow and improve doesn't mean you have to hate the current version of yourself. You may do things that you regret or act in ways you didn't mean to, but constantly being critical of yourself isn't going to help you work towards positive change. Instead, you'll need to embrace positivity and, ultimately, self-love.

Self-love is the practice of caring for yourself, showing yourself respect, and recognising all the wonderful parts of yourself. It can be tough to embrace self-love if you have low self-esteem, especially if anxious and depressed thoughts work to undermine your positive mindset. You may struggle to see to your own needs, and there may be times where you have difficulty thinking of yourself in a positive light. This is all the more reason to embrace self-love, as it will help you understand that there is plenty to celebrate about you. Identify your positive traits like empathy, honesty, compassion, loyalty, and diligence, and remind yourself that you are worthy of love and appreciation from both yourself and the people in your life.

Contrary to popular belief, practicing self-love doesn't mean you can never accept that there are things about yourself you'd like to improve. In fact, self-love can be a powerful positive motivator for change, as the more you appreciate your positive qualities, the easier it becomes to turn those positive qualities into a strong foundation upon which you can construct the new you. Focus on what you love, and refuse to be cruel to yourself over the aspects you'd like to improve. Accept that change takes time, and that you are worthy of the effort you're putting into self-improvement. As you continue to build up your mental strength and address the struggles you've been facing, always remind yourself that you are more courageous and powerful than you know. This will help you keep your head above the water even in highly stressful situations.

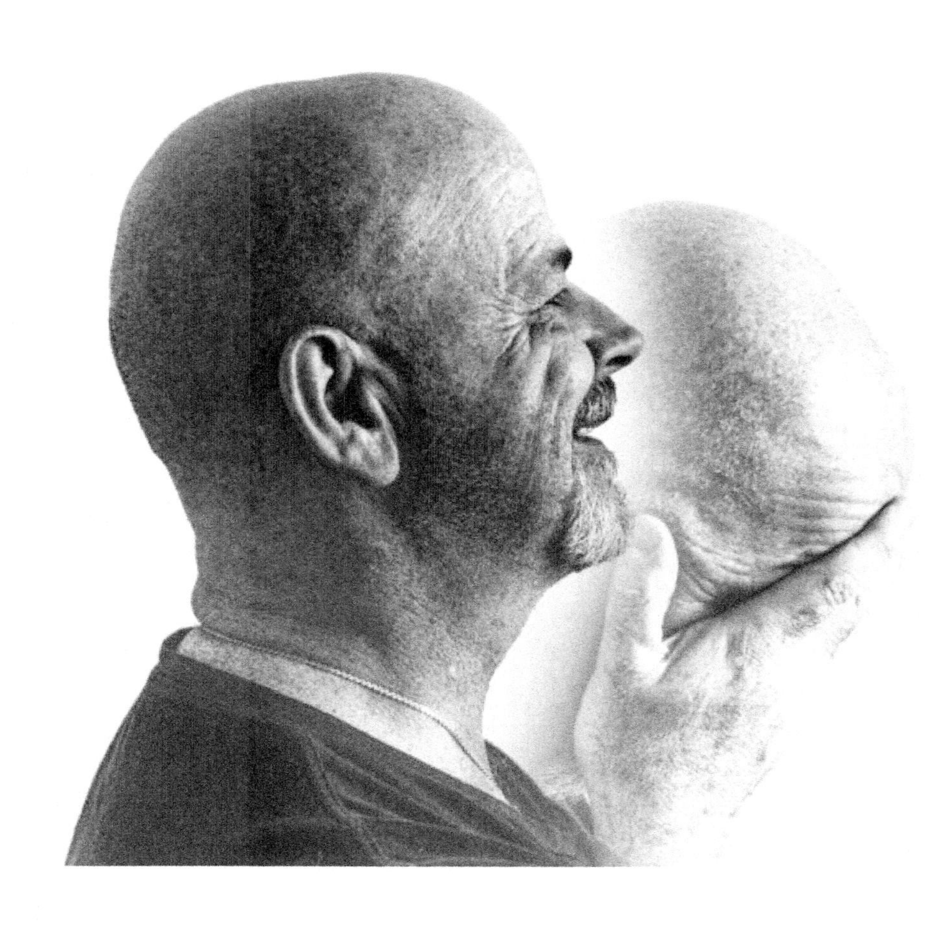

Chapter 3: How to Manage Yourself With Others

Social uncertainty and anxiety are incredibly common. Many people worry about whether or not they're saying or doing the right things in social situations, or if they're coming across as likeable and approachable. If this sounds familiar, you might have also noticed that there are times where you're unusually withdrawn in conversations, or just the opposite—maybe you attempt to be outgoing, but it falls flat because you're not comfortable, nor are you being true to yourself. If you say something that doesn't get a positive reception, this can feel like a personal failure, and you may start thinking critically about yourself, assuming that friends and strangers alike don't enjoy your company. It's natural to want others to like you, but if you put too much stock in others' opinions, you may accidentally begin to undermine your opinion of yourself.

Navigating social situations can be tricky if the very idea makes your heartbeat a little faster. However, it's a necessary skill for everyone to learn, even and especially if you often find yourself anxious around others. In fact, avoiding social situations can actually make them more difficult to navigate, as they become more frightening in your mind. This process, known as avoidance, doesn't do you any favours. For one, they can lead to the loss of good opportunities, as you don't take advantage of ways to improve your social life or professional standing. Maybe taking over that presentation your boss wants you to do would make you a good candidate for a promotion, but if you're too nervous to accept, the chance will simply pass you by.

Additionally, if your experiences with something you find frightening are limited to negative ones, refusing to ever give it another shot means these negative experiences become your entire frame of reference. Let's say you ask someone out on a date and they reject you harshly. If you let this convince you to stop asking people out when you're interested in them, the only experience you have with dating will be rejection. It's nearly impossible to regain your confidence in this case, since every time you think about trying again, you remember the single negative experience. However, let's say that instead, you're willing to try again shortly after the first rejection because you haven't given yourself to ruminate and turn one bad experience into a bogeyman. Most people won't reject you harshly—they might say no kindly, or they might even say yes. The more experience you get, the easier it is to see the situation in a logical and fair light, rather than relying on a single experience. Ditching avoidance will help you regain your confidence in social situations and push back against anxious thoughts that encourage you to always assume the worst possible outcome. The more practice you get, the less likely you are to catastrophise.

You can make this process of gradually reacclimating yourself to the things that frighten you a bit easier by focusing on other skills like communication and confidence. If you can build up these social skills, you'll have a better grasp on social cues, and you'll have less difficulty navigating different conversations with ease. As you get more practice, you may find that you're quite a bit more extroverted than you once thought! But before you can accurately decide how comfortable you are in social situations, you must first improve your communication skills.

Effective Communication

Good communication allows us to understand others and to be understood ourselves. While we can all hold a basic conversation, it's very easy for our messages to be misunderstood if we're not communicating clearly. It's equally as easy to misunderstand others if we're not paying proper attention to what they have to say. Poor communication can result in awkwardness, hurt feelings, and unintended consequences, so improving your communication skills is a worthwhile endeavour for anyone.

Great communicators tend to experience less social anxiety, as they're not so worried that they'll come off the wrong way. They know how to make others understand them, and they have enough self-confidence to not be too put-out if the person they're talking to doesn't immediately take a liking to them. One important thing to remember if you struggle with social anxiety is that it's okay if you don't come off in the best light in every single conversation. Not everyone will like you, and that doesn't mean you're a bad person or that you're unworthy of being loved. Even the most beloved celebrity is hated by some people, so why should you have to live up to a higher standard? As you start to embrace the idea that you don't always have to bend over backwards to make yourself more likeable, you'll feel more confident and comfortable sharing your genuine opinions rather than trying to constantly assess social situations and react in ways you think others will like. You'll start to be more genuine, and you'll build up your sense of self-assurance in time. This will help you let go of stress fuelled by the desire to always be liked and to live up to others' standards. Instead, you'll be able to start living for yourself.

Achieving this level of social confidence requires practice, of course. You're not going to do your social skills any favours by avoiding social outings whenever you can. Try to embrace every opportunity for getting in some more practice in social situations, especially at casual gatherings where there are low stakes and you don't need to worry so much about being liked. Over time, you'll learn how to communicate more effectively and make yourself clear, even when you're in more stressful situations.

Being Understood

Effective communication is a skill that even the most outgoing people may struggle with if they don't put time and effort into developing it. This is in part because it relies heavily on emotional intelligence, which is often overlooked in schools in favour of cognitive abilities. The truth is that no matter how smart you are, if you can't communicate your ideas to others, your ability to remain mentally strong is going to be limited, and you'll likely feel more tension in social situations.

Good communication skills are based on both your ability to be clear and your ability to listen to others. First, consider the message you want to send, as well as your audience. If you're at work, you'll need to use professional language and remain respectful of your bosses and co-workers. If you're in a more casual setting, such as among friends, you might not have to choose your words quite so carefully. Additionally, different people will react to the same information in different ways, so take this into account when you talk to others. Practice your listening and understanding skills as well. You want to talk with others, not at them, and this means paying attention to what they're saying so you can respond appropriately. The more adept you become at navigating conversations and getting your meaning across, the easier it will be to let go of anxious thoughts and feelings that might otherwise hamper communication.

Avoid Putting Yourself Down

One common mistake many people dealing with anxiety make is putting themselves down in conversations. While it's important to pay attention to the needs and viewpoints of others so you can communicate with them effectively, this doesn't mean you have to hold them in higher regard than yourself. In fact, constantly putting yourself down can be actively harmful for your mental health, even if you're doing it as a joke. Negative self-talk is a form of self-sabotage, and you need to drop this bad habit before it can do any more harm to your well-being.

How you think and talk about yourself impacts how you feel about yourself. If you're constantly joking about how stupid you were to make a mistake, or how useless or disappointing you are, these ideas don't stay jokes for very long. After all, you likely wouldn't put yourself down quite so much if there wasn't a part of you that believed these things to some degree. The more often you repeat these negative sentiments, the more you reinforce these ideas, and the more acceptable it becomes for you to think of yourself in a negative light. If you want to improve your mental health, you need to stop these practices and treat yourself kindly and with dignity.

Think of how you would talk about your friends and whether or not they would appreciate being spoken about this way. Would you find it acceptable to say your friend was useless at their job, or that they're an idiot? Of course not. So, why is it any more acceptable to say these things about yourself? Instead of constantly speaking negatively about your failings, try to recognise your accomplishments, just like you would for your friends. Congratulate yourself on the things you did well. Talk yourself up. If you don't feel comfortable offering yourself genuine compliments just yet, you can do this jokingly too. Try saying things like, "Nobody's better than me!" or "Everybody in this building wants to be my friend." Talking yourself up jokingly can be just as effective as trying to pay yourself genuine compliments. The same rule about insults applies here too; what starts as a joke quickly becomes sincere if you say it enough times. Praise yourself when you can, and avoid saying unkind, untrue things about yourself that you would never say to anyone else.

Another way you might put yourself down is by consistently devaluing your ideas in conversations. You might preface your sentences with, "It might just be me, but..." or "This might be stupid, but..." even though these caveats should be avoided. These sort of phrases undermine your confidence in yourself, and they make it harder for others to take you seriously too. You probably wouldn't give an idea serious consideration if the person saying it didn't think it was a good idea, so others might dismiss your thoughts and feelings out of hand if you're not confident and assertive when you share them. You don't have to talk over others to be more assertive, but you do have to really believe in what you're saying, and that means simply saying it without any self-doubt. As you learn to stop devaluing your ideas, others will find the value in them as well, and they'll be more open to listening to your input.

Make Yourself Heard

Consider this scenario: You're working as hard as you can at your job, but it seems like your boss never notices you. Nothing you do seems to get his attention, and you're feeling like nobody recognises all the work you do. You go home at the end of the day and you want to do something nice, so you decide to make dinner for your partner. At dinner, your partner eats and chats with you, and while they clearly like the meal, they don't explicitly say thank you or compliment your cooking. Incensed, you perceive this as your efforts being looked over once again, and you either start an argument because of this slight or you stew unhappily, growing even more resentful.

The mistake in this scenario was poor communication. Your partner had no way of knowing you were desperate for your efforts to be recognised, so they didn't understand when you suddenly got mad at them. Whether you close yourself off or blow up and start yelling, you're not communicating this need to them, so they have no opportunity to help fulfil it. While it would be nice if everyone could react exactly how we'd like them to, this just isn't realistic. Sometimes, we have to talk about things that might make us sound needy or which might be embarrassing for us to admit, because otherwise there's no way for others to know that we need some extra support.

Clear communication will help you navigate relationships and strengthen your bonds with your loved ones, even during moments of conflict. It's necessary to make yourself heard by explaining your needs and feelings. Rather than jumping right to yelling, discuss why you're angry and what made you upset, and be willing to listen to others' reasoning about why they act and feel the way they do too. Make it clear why certain situations are more upsetting or frustrating for you so others can take more care when broaching these subjects. This won't prevent all conflicts, but it will help you resolve conflicts in a less explosive manner where no one carries lingering resentment after the fact. If you speak up about your feelings, others can be more aware of them, and you can be more aware of their feelings in return.

Developing Your Social Self-Confidence

We've mentioned the need for confidence, but how exactly do you go about becoming more confident? First, it is necessary to remember that confidence is something you must practice before you really start to feel it. Even if you really don't feel confident at all, projecting a more self-assured version of yourself can help you feel more comfortable. This means both speaking confidently and having your body language reflect this. It is okay to feel awkward or uncertain at first. This is only natural if you have spent a long time feeling unsure of yourself and trying to take up as little space as possible. The more you practice projecting confidence, the more confident you will feel.

Projecting Confidence

You might have heard the phrase "fake it till you make it" before. This means that if you pretend to be the person you want to become, adopting the right mannerisms and other behaviours, you will eventually become that person. This is often applied to the journey of becoming more confident. In short, if you pretend you're someone who has a lot of confidence, purposefully acting as self-assured, outgoing people do, others will treat you like you really are confident, and you'll start to really become more confident too.

To project confidence, you'll need to know what behaviours to replicate. Think of a friend or family member who is very outgoing and certain of themselves. What do they do in social situations? How do they speak to others? Are they quiet or loud? Do they wait hesitantly until everyone else has stopped talking to timidly make their point, or do they find opportunities to make themselves heard without stepping on others' toes? Do they run from conflict and avoid speaking their mind, or do they ensure their opinion is always considered and try to prove their point? Do they sit hunched over with their arms wrapped around themselves, or are they more relaxed and open with their posture? Once you know what confidence looks like from an outsider perspective, you can start to replicate it.

Of course, faking confidence is still just that: fake. It will take some time until you start to really feel the fruits of your labour, and during that time you should reinforce these confident behaviours with other methods that will boost your sense of self-worth. You can't maintain a fake sense of self-esteem forever, but you can work on improving your confidence in the background while you project a more self-assured front, and combining these habits will have more lasting effects. Positive self-talk, as mentioned previously, is one part of the confidence puzzle. So is visualising positive outcomes rather than fixating on negative ones.

This will help you convince yourself that stepping outside of your comfort zone can be a good thing. The more experience you can get with the things that make you nervous, the less power this kind of fear will have over you, and the easier it will be for you to embrace a more positive mindset and cultivate real confidence.

Power Poses

Body language is an incredibly important part of confidence. Posture that is more relaxed and easy-going communicates your confidence to others, but it can also help you feel more confident too. Adopting "power poses," where you stand up tall with your shoulders back and your chin out, helps you radiate confidence both outwardly and inwardly.

Power poses became popular after a TED talk by social psychologist Amy Cuddy, where she shared her findings on the different effects of expansive and contracted poses on mentality. Expansive poses involved the test subjects making themselves taller, taking up more space, and generally seeming more open and comfortable. Contracted poses meant participants were instructed to hunch in on themselves, cross their legs, and otherwise make themselves smaller. Through the study, "researchers found that after adopting an expansive pose, study participants felt more powerful, took more risk in a gambling task and performed better in a mock interview than those who had adopted contracted poses" (Elsesser, 2020, para. 2). In other words, altering your posture can, in turn, alter how you see yourself, as well as how others see you. If you need a little confidence boost, try standing up tall and allowing your body to reflect your desired confidence level. You might just find that this gives you the certainty you need to speak your mind and improve your social skills.

Once again, power posing is just one part of developing a more confident mindset. You'll need to reinforce these poses with real work to change negative thought patterns and to start holding yourself in higher regard. But they can still be a useful tool, and their benefits shouldn't be overlooked.

Rooting out Toxicity

Becoming more socially confident is just as much about whom you talk to and surround yourself with as it is about how you yourself act and speak. Your social circle is a significant contributor to your mindset. If you surround yourself with helpful, supportive people who want to see you succeed and who are willing to lend a hand when you need one, you'll find it easier to step out of your comfort zone and lower your emotional walls. On the other hand, if the people you talk to always seem to put you down, make you feel small, or discourage you from putting yourself out there and trying new things, you're going to find it very difficult to break free of the chains holding you back from self-improvement.

In some cases, a few offhand comments turn into a pattern of behaviour that becomes toxic. You might have people in your life who are draining to be around and who frequently make you believe that you can never improve your life or yourself. Maybe they always put a negative spin on otherwise positive events. Maybe they encourage you to put off work and to waste time instead, even if it puts your job performance in jeopardy. Maybe they frequently belittle you and make your accomplishments feel worthless. If someone in your life isn't supportive of your goals or the person you're trying to become, it may not be worth keeping them in your life any longer, regardless of how long you've known them. The truth is that you cannot grow and flourish in a toxic environment. Even if it can be hard to address these issues and end long-standing relationships, rooting out and removing toxicity paves the road to self-improvement and social confidence.

Understanding Toxic Relationships

Toxic relationships are characterised by repeated, harmful behaviours from one or both parties that cannot be reconciled. These behaviours may be emotionally or physically harmful, or both. It's important to note that one or two comments do not typically constitute a toxic relationship. For example, if your boss tells you that you missed a deadline and you need to take your work more seriously, this alone isn't toxic. However, if they consistently berate you and put you down, or if they purposefully make your job harder or ostracise you from feeling involved with the company, this is often a sign of a toxic working environment.

Toxic relationships are most easily understood through the lens of their effect on you. If you're in a toxic relationship, be it at the workplace, among friends, with a romantic partner, or within your family, good indicators include feeling belittled, talked down to, and made to feel insecure. You may not feel entirely safe around them. In some cases, toxic relationships can become abusive when someone tries to take control over your life, deciding what you are and aren't allowed to do, but not all toxic relationships are this severe. It's also important to remember that some dysfunctional relationships aren't toxic on purpose, and these issues can arise from conflicting personalities or from unresolved issues on behalf of the toxic person. While someone who is toxic may genuinely care for you, that doesn't give them the right to put you down, and you need to separate yourself from them for your own good.

Removing Toxicity

To remove yourself from toxicity in all areas of your life, you must first decide if the issue making a given relationship toxic is something that can be improved or if it is safer for you to end the relationship. For example, you might feel like your workplace is toxic because it monopolises your time and encourages a culture of 'workaholism' that negatively impacts your mental health. Sometimes, you can resolve this issue just by discussing it with your boss or a supervisor. They may not realise the trouble they're putting you through, and bringing up the issues you've been facing can mean you can now work together to find ways to ease the burden your work places on your mental health.

In other cases, however, the other party may be unwilling or unable to work with you to end the toxic component of the relationship. In this case, it is necessary to put yourself first and leave the relationship for your own good. This can be incredibly difficult. You might have to give up a well-paying job if it's putting too much stress on your mental and physical health. You may have to remove a friend or family member from your life who you were very close to many years ago, but who has since become someone who puts you down or holds your back. It may seem impossible to end these kinds of relationships, but once you do, you will start to regain some of your lost confidence, and you'll be ready to move forward with your life.

Developing a Support Network

Ridding yourself of toxic relationships is much easier when you can replace them with a group of supportive, caring people who are excited to help you along your journey to self-improvement. It's critical to develop a strong support network so you can bolster your mental health and well-being with the love and care of the people you trust.

Your support network can be made up of people from many different areas of your life. They can be your family members, as you've known them the longest, and you always have that familiar bond tying you together. They can be your friends, as these are the people you have decided to spend your time with, and they have stood by you through thick and thin. They can be your co-workers and peers, who may motivate you to excel at work and continue to pursue improvements in your career. They can even be people you meet through support groups or in online communities, who want to foster a kind and caring environment even though you might not have known them previously, and even if you never meet in person. Supportive people can come from anywhere. All you need to do is identify people who genuinely care about you and who want to help you see the best in yourself.

Surrounding yourself with supportive people is the first step in making your environment more conducive to improving your mental health. Of course, your environment includes much more than the people in your life. Most notably, it includes the things you choose to surround yourself with inside your home, where you should be able to feel most comfortable.

Chapter 4: How to Create the Best Environment

Your home is your base of operations. It should function as a safe, comfortable space where you can relax and be yourself, serving as your sanctuary in a world that can often test the boundaries of your comfort zone. While it's good to occasionally have experiences that are unusual and encourage you to embrace new things, you should always have somewhere to return to where you don't feel the pressures of the outside world. This is the best way to improve your mental fitness without overexerting yourself—even the best athletes understand the necessity for recovery time, and the same idea that applies to physical strength training is just as applicable for mental strength training.

Unfortunately, you cannot fully consider your home to be your sanctuary if it is in a constant state of disarray. If your home is just as hectic as the outside world, it's not going to be much of a reprieve. This chaos can come in many different forms. It might arise from a complicated relationship with other members of the household, especially if you don't have a dedicated space you can call your own where you can retreat to for alone time. It can come from having a lot of clutter and visual noise in the house. It may be the result of different parts of your home serving as reminders of tasks you need to get done or unhappy memories full of regret, which ensure you can never fully turn your brain off. No matter what form this disarray takes, it can seriously interfere with the restorative role your home should play in your journey towards greater mental fitness.

Every creature on earth needs the right environment to thrive. A fish cannot survive on land, a cold-blooded snake cannot thrive in the tundra, and you cannot achieve your full potential if you are constantly stressed out by your surroundings. Organising your home, with both physical clean-up and mental or emotional decluttering, will help you create an environment that is conducive to true rest and relaxation. If you can eliminate sources of stress and worry from your home, or even just a single room, you can extend these feelings of safety and security to situations outside the home, always knowing you have somewhere you can completely destress without fear. Altering your environment is one of the best things you can do for yourself to improve your mental health, and in a world that grows more fast-paced and chaotic by the day, making even a few adjustments to your sanctuary can make a world of difference.

How Your Environment Affects Your Mentality

Our brains react differently depending on the space we're in. For example, when you enter your kitchen, you might feel hungrier right away because your brain recognises it as the place you eat your meals. If you have a dedicated study in your home, you might have an easier time focusing there than you would if you tried to work in your bedroom, where you would likely only feel tired. If you've ever heard of Pavlovian conditioning, this phenomenon should sound familiar to you. Essentially, it is the idea that we learn to associate certain stimuli, such as the rooms we're in, with the tasks we perform after encountering those stimuli. The more we condition ourselves to associate the kitchen with food and the study with work, the easier it is for us to mentally prepare ourselves for these tasks in their dedicated environments.

This can work against us too, if we're not careful. For example, let's say you try to do some work while sitting on your bed. In addition to feeling more tired while you're working, doing such an unusual task in the wrong environment actually weakens your mental association between being in bed and falling asleep. Next time you lay down, you'll have a harder time turning your brain off and getting some rest, because part of you is ready to work instead. This can lead to insomnia, which in turn results in bad moods and a limited ability to focus throughout the day. For this reason, it's best to designate certain areas for different tasks and to stick to these associations without deviation whenever possible.

Sometimes, changing up our environment can help us feel more motivated to complete a tedious task. A little variety does wonders for peaking our interest, so we might try to work outside if we're feeling a little too cooped up. But this can just as easily become distracting and make it harder to get anything done if the environment isn't well suited to the task at hand. This is just one reason why it's so important to consider the role your environment plays in your productivity, motivation, and mindset.

Clutter Versus Focus

In recent years, there has been a strong push towards minimalism, which is the practice of reducing the number of objects in your home to bare essentials and ensuring everything has a proper place in your house. Minimalism spiked in popularity with the release of the show *Tidying Up with Marie Kondo* in 2019, though many people were already living a less cluttered lifestyle long before this, and with good reason. Whether or not you're willing to go completely minimalist, there is some scientific basis to the idea that too much clutter can sour your mood and lead to a more negative and easily-distracted mindset, interfering with productivity and happiness.

Clutter has a much more worrisome impact than simply not being able to find that shirt that's hidden at the back of your wardrobe. It has been strongly linked to higher levels of the stress hormone cortisol, such as in a 2010 study published in the *Personality & Social Psychology Bulletin,* which found that women "with higher stressful home scores had flatter diurnal slopes of cortisol, a profile associated with adverse health outcomes, whereas women with higher restorative home scores had steeper cortisol slopes." The researchers for this study also found that "women with higher stressful home scores had increased depressed mood over the course of the day, whereas women with higher restorative home scores had decreased depressed mood over the day" (Saxbe & Repetti, 2010, p. 71). This means that having higher amounts of clutter in your home is directly correlated with being more stressed throughout the day. This may in part be because when your home is full of clutter, your brain can't fully relax since there is so much visual noise for it to process. Without that relaxation period, you aren't reducing your stress levels, which contributes to worry, frustration, and poor moods.

Significant amounts of clutter can also make it harder to focus, whether you're directing that focus to working or relaxing. When you're surrounded by a large number of things, it's very easy to get distracted from whatever task is at hand. Your brain is constantly processing visual information, which takes up a lot of its available attention and energy. Even if you really try to buckle down and focus, you might find your eyes drifting to distractions in your environment, especially if these distractions seem like more fun than whatever you're currently doing. This can seriously harm your productivity if your office at work is full of clutter, but it's a problem at home too. If you're trying to practice a hobby, you won't get nearly as much out of your efforts if you keep taking breaks to check out something else. Even if your goal is to relax, take a nice bubble bath, and watch some TV, reminders of other tasks you need to complete that are within your line of sight can make this break much less restful. If you really want to sharpen your focus, one of the best things you can do for yourself is remove anything that gets in the way of the task you're trying to complete, and that means cutting back on excessive clutter and rearranging your environment to make the task as easy as possible.

Your Surroundings and Your Physical Health

Even though the connection may not seem clear at first, a home that's in disarray can subconsciously encourage poor eating habits as well. A 2016 study of Australian and U.S. students published in the journal *Environment and Behavior* found that "college students were twice as likely to reach for sugar-rich foods when they were stressed in a messy kitchen" (Truong, 2019, para. 9). While your mindset can help defend against the allure of these temptations, the chaotic environment increases stress and the desire for a quick mood boost to alleviate that stress, which can be readily found in the form of junk food. If you're constantly living in a busy, cluttered environment, it will become more and more difficult to ignore these cravings. As you know, healthy eating supports a healthy brain, so this can be a serious problem for both your body and your mind.

All in all, clutter is something you definitely want to avoid if you're looking to improve your mental fitness. When you construct your environment to support rest and focus, you'll find it far easier to live a more restful, more focused life. Turn your home into somewhere you feel completely comfortable and relaxed by cutting out clutter and crafting the perfect zone for your needs.

Reducing Chaos

Excessive amounts of chaos in your environment forces your brain to work overtime, which is exactly what you don't want in your home sanctuary. To combat this, you'll want to start reducing the amount of visual noise within your house. The first step is to get rid of everything you don't need and which only serves to make you feel crowded and uncomfortable. This also means cutting back on how much junk you buy for your house on a regular basis, which will help you keep the clutter down.

Of course, there's much more to reducing chaos than just throwing things away. There's also the need for organisation in any space you're in, from the bedroom to the kitchen. You can also make smart decorating choices, such as paying attention to the ways different colours and lights affect how welcoming a room is. You don't necessarily have to overhaul your entire style of interior decorating, nor do you have to throw away everything you own. But making an effort to eliminate any examples of significant visual chaos in your home is a simple and easy way you can ensure you have a safe, calm location where you can rest and relax no matter how stressful the rest of your day has been or will be.

Do Some Spring Cleaning

A great deal of what makes the average home rather chaotic is just how much stuff there is inside. Most people own far more than they need, to the point where you may rarely or even never use half of the things inside your home. Try this exercise right now: walk to your wardrobe and look at all your shirts. How many of them have you worn in the last year? Can you remember the last time you wore them? Are there any shirts you bought at least a few months ago that still have the tags on them? If so, you're probably buying more than you would ever reasonably use, and this kind of clutter is likely present in many other areas of the house.

The idea of spring cleaning might sound like an arduous process you just don't have time for if you live a busy life, but if you break it down into smaller tasks you can complete over a week or two, you can really make a difference in your everyday quality of life. Start with just one room. Pick a room where you want to feel the most comfortable and relaxed, such as your bedroom. Begin with something small, like clearing away any trash in the room. Collect used cups, plates, napkins, cans, and anything else you've been meaning to clean up for a while and wash or toss them. This is an incredibly tiny step that you can do in just a few minutes, so you have no excuse for continuing to put it off. By actually doing it rather than continuing to procrastinate, you've taken the first step in decluttering and started to build your momentum.

Next comes the hard part: getting rid of things you no longer need. This requires a critical eye and a willingness to part with things you just aren't using. You don't have to turn your bedroom into a completely minimalist space, but you need to be honest with yourself about whether you're actually getting any benefit out of something or if you're just holding onto it because you feel guilty about the idea of throwing it away. If a decoration that would otherwise be 'useless' genuinely makes you happy to have around, by all means keep it, but only keep what you really like and get rid of all other knick-knacks. Sort through your clothes and figure out what still fits you and what you still like, then set everything else aside. Clean out desk drawers, side tables, shoe racks, shelves for storing games or books, and anything else in your room with the mindset that you are only going to keep the things you need and that you use frequently enough to justify their continued presence in your life.

If you've got a few things that ride the line between 'keep' and 'toss' and you're not sure what to do with them, try collecting them all in a bin and keeping them under the bed or in a cupboard for about a month. If you open the bin and use something from it, you can return that item to its usual place and keep it. If you still haven't used it after a month, there's a good chance you don't really need it after all, and you can get rid of it without feeling too bad. If you feel a little guilty about tossing so much stuff, look for alternate ways to get rid of these items where they won't end up in a landfill. For example, you can donate blankets, clothes, and toys as long as they're in good condition, and clothing scraps can be repurposed into rags for washing the car or crafty projects. You might give some of your belongings to friends and family, who could make better use of them. Recycle what you can, and only throw away what you absolutely have to.

From here, you can continue breaking the work down into small chunks and completing each chunk one after another over the span of a few days. Make sure all clothes are put away if they're clean or moved to the laundry basket if they're dirty. Straighten your desk drawers and move any stray items back to where they belong. Now that you have less to work with, it should be easier to find a place for everything, and your room should feel significantly less cluttered. You can then repeat this practice with every other room in your house until you've decluttered your entire living space. If you live with other people, just make sure you don't get rid of anything they might be using. You might find it more productive to include them in this decluttering exercise, as the work will go by faster and you'll have a clear idea of what everyone does and doesn't want to keep.

Create Dedicated Spaces

One of the most significant barriers to staying focused throughout the day is having a lot of distractions in your environment. Decluttering can help you get rid of some of these, but others are the product of trying to do too many different kinds of tasks in the same space. For example, let's say you want to work on a creative hobby like drawing or writing, but you have a TV or video game console on or near the same desk. Maybe you want to get some work done at home, but you decide to do it while sitting on your bed. In these cases, your brain is getting mixed messages because the environment doesn't support the activity you're doing. You're introducing too many potential distractions and weakening the mental connections between your environment and your task, making it harder to focus on any one thing. Before long, you'll find yourself turning your attention away from your art project and towards the TV, or if you're trying to work in bed, your eyelids will start to droop. In either case, the problem arises because you didn't dedicate specific spaces to different tasks.

The need for dedicated spaces is even more apparent if you live in a flat with an open plan, where even the borders between different rooms might not be clearly defined. In any case, furnish these rooms in ways that make sense for the tasks you're completing in them, and move any distractions to their appropriate area in the house. If you work from home, clear away anything that might split your focus, like TV, video games, and books, and move these things to their own space. If you want to create a part of your house where you feel comfortable relaxing, try to close the area in a little to create a sense of a physical barrier between you and the stresses of the outside world, and rid the space of anything loud and disruptive. Fill it with things that will help you calm down and make you feel safe such as fluffy throw blankets, pillows, some reading material, and maybe a relaxing candle or two. The more you craft different areas of your home to suit your various needs, the easier it will be to reset your mind into distinct 'work' and 'play' modes, sharpening your focus and alleviating the tension caused by too many distractions.

Get Organised

Once you start clearing away the unnecessary clutter and turning different areas of your home into dedicated spaces, you can then move on to organisation. Being organised is about much more than just keeping your spaces tidy. It's about arranging items in a way that makes it easier to use the space for its intended purpose. For example, let's say you have a home office with a desk. You could organise the contents of the desk drawers by what kind of task they help you perform, so you don't have to go hunting for everything to complete a single objective. This will allow you to remain in a focused headspace while working on a task, where you won't have to break your concentration to figure out where things are.

The same rule applies to other areas of the home as well. You can cut a 45 minute meal down to 30 minutes just by making sure all your cookware is washed and stored in the right place all the time. Even spaces that are meant for relaxation should be organised in a way that facilitates ease of use, as hunting for that book you want to read or that lavender candle you want to burn can create unnecessary stress and anxiety, subconsciously discouraging you from fully turning off your brain and relaxing. If you have to enter a more chaotic room to find something you need to relax, you're interrupting your relaxation process. In this way, good organisation is absolutely key for allowing yourself to completely unwind in your home, just as much as it's necessary for staying focused too.

Consider Your Colour Choices and Lighting

Streamlining your environment for different purposes, such as work or relaxation, can also be more effective if you incorporate the science behind interior decorating. One of the most popular examples of this is colour theory, which suggests that the dominant colour of a room has an impact on how we feel when we're in the room. Colour has a psychological component to it because we associate different colours with different feelings. For example, if you wanted to make a room seem more relaxing, you'd likely stick to cooler colours like different hues of blue, purple, and dark green. Blue is especially calming, purple has long since become associated with royalty and, therefore, luxury, and green is a natural colour. Each of these colours can subconsciously encourage relaxation as a result of these associations. On the other hand, bright and lively colours like red, yellow, and pink denote passion and enthusiasm, which might be better for more playful rooms. Using red in the kitchen is even believed to encourage hunger, so it's a common colour for pots and pans. You don't necessarily have to repaint your entire house, but accenting rooms with appropriate colour choices can help reinforce the type of environment you're trying to create.

Lighting can have an equally strong effect on making a room feel more welcoming. A well-lit room is naturally more approachable than a dark one, as you can see everything inside and there are no shadowy corners that might convince your brain there's a reason to be wary. However, harsh bright lights might be more alienating, washing out the colour in a room and making our minds think of hospitals and other clinical areas that are less than relaxing. If you want to promote rest and eliminate anxiety, go for warmer coloured lights strategically placed to illuminate all corners of a room.

Well-thought-out decorating choices can really reinforce the way you perceive a room, especially if you make a point to include elements that are pleasing to your senses. Photos and paintings give you something to look at while also giving a room character. The effect of a large landscape is going to be very different from an average abstract piece, so decide what kind of message these decorations give off before incorporating them into a room. You might also incorporate pleasant, calming smells like lavender and sage in the form of candles or incense burners. Each of your decorating choices should help you maintain a calming atmosphere, and there's no better way to do this than cutting back on clutter and using what remains to appeal to your senses.

Incorporate Nature

Nature has a scientifically-proven positive effect on your mentality. Spending more time in nature can quiet constant buzzing thoughts that contribute to worry and rumination, improving your mental health. If you can't get outside or you just want to make your home a little more restful, you can experience the stress reduction that comes with being in nature from bringing natural elements indoors. This includes potted plants, natural textiles like wood and stone, and plant or animal prints on accent furniture.

Nature's benefits have also been reported to come from ambient noises. The results of a 2017 research study published in *Scientific Reports* and later recorded in Harvard Men's Health Watch found that "listening to natural sounds caused the listeners' brain connectivity to reflect an outward-directed focus of attention, a process that occurs during wakeful rest periods like daydreaming," while "listening to artificial sounds created an inward-directed focus, which occurs during states of anxiety, post-traumatic stress disorder, and depression" (Harvard Men's Health Watch, 2021, para. 9). When you feel like you're immersed in nature, even if you're just listening to bird calls or the sound of a babbling stream, you really do feel more relaxed. Adding these elements to your home in the form of recordings, tabletop fountains, and even cracking open a window to let some noise and light in is a perfect way to promote natural rejuvenation while remaining in your comfort zone.

Creating Your Comfort Zone

While you can do a great deal to ensure your entire home is more welcoming and better at promoting rest, you may not have the luxury of being able to make big changes to your whole house. Maybe you live with roommates, who have spaces of their own and who may not approve of certain renovations to common areas. Maybe you have a partner and kids, or a parent staying with you. It's hard to find peace in a large household like these, and nearly impossible to keep everything decluttered and organised if your kids are constantly rummaging through it all. In these cases, it's still important to create a comfort zone that is exclusively for your rest and relaxation.

Your comfort zone should be somewhere you can go to be alone. It should serve as a retreat from the rest of the world, giving you a calm and quiet space to recover, destress, and be entirely yourself without the fear of being judged by others. Your comfort zone could be a single room or even a corner you've laid out as your own personal area. The most important rule of creating your comfort zone is that it must be entirely your space, not a space for your kids, spouse, roommates, or parents.

Once you have established the parameters, you can start personalising your comfort zone so it lines up with your individual values and needs. Looking for somewhere quiet to get some reading done? Need somewhere you can feel completely at peace? Want to do some meditation or just sneak in a quick nap? Your comfort zone should give you the space and privacy you need to fully relax so you can feel refreshed and ready to take on all of life's difficulties when you emerge.

If your comfort zone is just one part of a room, try putting up some dividers or otherwise enclosing the area to give yourself your own space. You can reinforce this sense of privacy with noise-cancelling headphones and plenty of comfortable seating complete with pillows and blankets. Your goal should be creating your little piece of paradise, so stock it with all the things you need to feel fully relaxed, and leave out anything that causes stress.

As you start to feel fully relaxed, your thoughts will stop racing, and you'll be able to get the rest you need to feel mentally fit when you leave your comfort zone. Creating a relaxing environment in your home is absolutely key to building up your mental strength, so never ignore the effect your surroundings have on your mood.

freedom
DARK SIDE
SOCIAL
SKILLS
FIRED!!!
DRUGS $
BORDERLINE PERSONALITY DISORDER
BIPOLAR DISORDER
SELF ESTEEM
MONEY
SEASONAL AFFECTIVE DISORDER
LOVE
1♡
AFFECTION
DEVOTION
LOVE
@
REPOST
LIKE
APPRECIATION
OCD
PTSD
FEAR

Chapter 5: How to Deal With Immediate and Long-Term Stress and Calm Your Anxiety

There are countless sources of stress you might encounter in your daily life. Stress can be understood in two different forms: immediate and long term. Immediate stress is something that creates a lot of anxiety in the moment. For example, if you have a fear of public speaking, being put on the spot to answer a question or give a presentation can cause immediate stress. You can often feel the sudden spike in your heart rate, and you may feel short of breath or experience other physical symptoms of anxiety if you can't bring your emotions back under control.

In many cases, stressful situations that cause immediate fear and worry can lead to long-term stress if they're not appropriately addressed. Long-term stress can also come from having too many responsibilities on your plate at once, meaning you're unable to complete any of them as efficiently as you would like. Long-term stress can be even more troubling than immediate stress, as it represents a negative pattern of anxiety, self-doubt, and fear that infiltrates many other areas of your life. Being constantly stressed out takes a toll on your body, as it can raise your blood pressure and lead to hypertension and related health conditions like a higher risk of having a heart attack or stroke. You may also have more trouble falling asleep as stress and anxiety keep you awake, and you might struggle to eat a nutritious diet as well. You will likely find it harder to concentrate if you've been under a lot of stress for a long time, making it even more difficult to manage the different responsibilities in your life causing this kind of stress.

No matter what form stress takes, it's important to address it as quickly as possible. The ideal method for doing so is a little different for everyone, but for the most part, stress tends to come from having too much to deal with at once, trying and failing to control everything in your life, and fearing the idea of branching out of your comfort zone. You can address each of these issues by shifting your mindset, starting with getting a little more comfortable with feelings of uncertainty.

Embracing Uncertainty

If you're a perfectionist, or if you fear the idea of being berated for not completely excelling at a project on your first try, there's a good chance that a significant portion of your stress comes from your inability to embrace uncertainty. You're terrified of the chance that you might mess up, and on some level, you feel that not being absolutely flawless at whatever you do will make others like you less. As a result, you choose to remain firmly in your comfort zone, and the idea of leaving it fills you with anxiety. While having a comfort zone to retreat to when times are tough can be extremely helpful, as you saw in the previous chapter, you don't want to spend your entire life there. If you do, you'll never learn new things, and you'll let countless opportunities pass you by.

Living life to its fullest means trying new things, no matter how uncertain they might make you. No one is a natural-born master the first time they attempt a new skill. Even the best people in their respective fields started out as novices, and they were only able to gain the skills they now have because they were willing to accept the possibility that they might fail a few times before they got better at their craft. If Michelangelo was so worried about the possibility of making a mistake, he would never have been able to sculpt using such a permanent medium as marble, and we wouldn't have his statue of David. If he was never willing to give painting a try alongside sculpting, we wouldn't have his work on the Sistine Chapel. This is an extreme example, of course—you don't have to develop Michelangelo-level skills to feel like you've succeeded at any task—but it showcases the point that if you allow your uncertainty to hold you back, you'll never give yourself the opportunity to try new things. If you're open to the idea of trying something that scares you a little at first, you just might find that with practice, you'll become incredible at something that you might have otherwise completely written off.

Developing a Growth Mindset

In life, we are often encouraged to pursue the things we are 'naturally' good at and ignore all the things we aren't. Think back to primary school and the ways teachers might have spoken about your grades. Maybe they looked at your poor math grades, shrugged, and said, "Well, you're not great at math, but you're doing well in history." If you struggle a little more than other kids in a certain subject, you might decide that subject just isn't for you, and never try to improve in it again. But what if those poor math grades were just because your teacher wasn't explaining the concepts to you in a way you could understand? What if you could have improved, if only you had been given the chance to work a little harder at something that didn't come easily to you rather than disregarding it entirely?

It's very easy to write off a skill forever if you don't show immediate promise in that area. This line of thinking comes from a fixed mindset, which suggests that you are predisposed to be good at some things and bad at others, and there's little you can do to change that. How smart you are, how good you are at creative pursuits, and even some elements of your character are immutable; you can improve upon your existing skills, but you cannot do much to make up for the areas where you find yourself lacking. Having a fixed mindset can be incredibly harmful to your ability to branch out from your already existing pool of skills. Attempting anything outside your wheelhouse is more likely to stress you out because you don't see any way you could ever become better at the things you're bad at. You learn to fear failure and avoid it at all costs, as failure becomes a confirmation that you lack the natural talent needed to succeed. You strive to constantly look smart, so you avoid anything that could lead others to perceive you as 'dumb' or incompetent, which means you never try anything that has a non-zero risk of failure. It is only by succeeding at what you are already talented at and ignoring everything else that you can maintain the perception of yourself as smart and competent, so you start avoiding anything that could challenge that belief.

These concerns start to fall away when you begin trading out your fixed mindset for a growth mindset. Having a growth mindset means embracing the idea that any skill can be learned with enough practice and effort. You are only limited by how you choose to spend your time, not by any inborn qualities that you can't change. In other words, you can excel at just about anything if you give yourself the opportunity to try, fail, pick yourself back up, and try again. Encouraging yourself to see the world through this growth mindset perspective can have incredibly positive results on your ability to embrace uncertainty. According to Carol Dweck, who largely pioneered the idea of the growth mindset as detailed in her 2006 book *Mindset: The New Psychology of Success*, found that "the belief that cherished qualities can be developed creates a passion for learning," (Dweck, 2006, p. 12) priming you to constantly try new things, undeterred by how good or bad your results are. When you have a growth mindset, you understand that you're not going to be good at everything right away, and that's okay. It's fine to mess up on your first few attempts as long as you use those attempts as a way to better understand what went wrong and what you can do to improve next time. Failure is part of the learning process, merely showing you where you could stand to improve, and it should be embraced, not avoided.

Redefining Failure

One reason you might be so unwilling to try new things is that they carry a high risk of failure. It's possible that for all your life, you were told that failure was to be avoided, and that being anything less than perfect was unacceptable. This idea may have been reinforced by a number of different sources. Maybe your parents praised you for top marks in your classes and took away dessert privileges when you got lower than a B. Maybe you had a professor in university who would call on students to answer questions they didn't know, attempting to use public humiliation as a teaching tool. Maybe you worked at a company where your boss would berate you for small slip-ups, to the point that you were too afraid to ask for help when you really needed it. These situations are all connected by a single idea: failure is bad and deserving of punishment, and only success will be tolerated. There is no room for learning something new because you'll have to fail before you fully grasp it.

If you've grown up constantly having your fear of failure reinforced, it's very hard to start thinking of it as anything other than the worst-case scenario. This makes it incredibly difficult to get back on your feet when you slip up, because you see these mistakes as personal, moral failings. Therefore, it's necessary to rethink how you view failure, encouraging yourself to view it as more of a learning opportunity. Nothing worth learning can be accomplished without making mistakes, but it is through those mistakes that you discover the most about the task at hand and about your own ability to improve. This will help you feel confident in situations that might have otherwise left you uncertain and discouraged you from trying something you might have enjoyed.

Expanding Your Comfort Zone

The best way to flip your view of failure on its head is to get more experience with it. This means trying something new. Ideally, you should start with something with very low stakes, such as a fun hobby you've always been interested in but you've never quite felt confident enough to try. You can make your first few attempts privately so you can start to internalise the idea that missing the mark on your first try is totally normal and no one is going to berate you for it. You don't have to force yourself to be the best you can be—you can simply enjoy the process of trying something new. Then, as you keep practicing the hobby, you'll see how your skills develop over time. Eventually, you'll feel more comfortable sharing your efforts, good or bad, with others, and you'll learn from your mistakes. If you mess up, you can always try again tomorrow.

Once you've built up this mindset with a low-stakes hobby, extending this concept to other areas of your life will become the next logical step. You'll feel confident in your ability to learn new things, and you won't feel quite as ashamed of your failures since you know you can recover from them, so you'll start actively embracing discomfort rather than avoiding it. This will, in turn, allow you to take advantage of the new opportunities that come your way. For example, the next time a new job opens up in a field you're interested in but where you don't have a lot of experience, you'll feel confident enough to apply, as the idea of learning on the job will no longer seem so frightening. Over time, you'll really start to feel comfortable in situations where the outcome is uncertain, seeing them as exciting challenges and chances to engage in self-improvement.

Focus on What You Can Control

If you find yourself frequently feeling anxious or stressed, it may be a result of trying to exert control over things you simply cannot change. You might have to deal with many things in life that you have very little control over. For example, maybe someone in your family gets sick or you lose your job. Try as you might, there's very little you can do to reverse these tragedies. You can't singlehandedly heal your family member, and you can't turn back the hands of time to prevent your boss from firing you. If you don't recognise and accept this, you're only going to hyperfocus on your inability to change the past and work miracles. This leaves you feeling powerless, and it can contribute to significant stress and frequent emotional outbursts.

Learning to recognise when things are out of your control is important because it allows you to shift your focus to the things you actually have power over, such as how you choose to react to these situations. If a family member is sick, you could respond to this information by moping and feeling sorry for them and yourself, but this doesn't help either of you feel any better about the situation. Instead, you can make an effort to lift both of your spirits a little by going to visit them, helping them manage responsibilities like childcare, or even just taking a moment to come to terms with the news so you can serve as an emotional rock for them. These are actions that are within your power, and they can make a real positive difference for both of you. Similarly, if you lose your job, it might be tempting to give up on work for a while, letting negative thoughts take over. However, it would be much more proactive if you took a moment to settle your thoughts, then put your energy into a job search. If you have enough savings, you could even see the situation as an opportunity to focus on practicing your hobbies and developing your marketable skills for a little while before you see where life takes you next.

No matter what happens in your life, you always have control over how you choose to respond to it, and this is where you will find the greatest sense of security in times of turmoil. Work to improve the things you can, like how you react to stressful situations, and learn to accept that there are some things you simply cannot change. Once you acknowledge this fact and start focusing on your control over yourself, you'll have a much easier time coping with difficult events.

Break Down Big Issues

Some tasks can feel especially overwhelming if you're looking at them in terms of all the work you'll need to do before you can say it's complete rather than taking things one step at a time. If a task seems too difficult when viewed from this perspective, you might avoid starting it entirely. You might convince yourself that you just don't have the time and energy to do everything right now, so it's better saved for a later date. Then you keep pushing it off over and over again. If the task has no deadline, you never make any progress on it, and it remains an imposing weight on your shoulders. If it has a deadline, all your procrastination will only make it even more difficult to complete everything before it's due, and you might end up missing the deadline completely. In either case, if you focus on how difficult the task seems when viewed in its entirety, your motivation will waver, and you won't make efficient use of your time.

The trick to turning big, stressful projects into completely manageable ones that fit comfortably into your daily schedule is to break them down into smaller chunks. Sure, the idea of cleaning your entire house might sound intimidating, but as you saw in the previous chapter, it becomes a lot more doable if you start with a single corner of a single room. Yes, developing a skill takes a lot of time and effort, but it is not nearly so difficult when you break it up into 20 or

30 minutes of dedicated practice each day. When you start seeing difficult tasks in terms of their smaller components, tackling these mini-tasks one at a time, you'll find that you can complete just about anything without all the stress that comes from procrastination and worry. The trick is to use the strategy of dividing and conquering to your advantage.

How to Divide and Conquer

It's all well and good to say that splitting up a task makes it easier to complete, but what does this mean in a practical sense? Is there a way to divide tasks so you can make progress on them even when you are completely lacking in motivation? To answer these questions, let's consider the example of cooking a meal. You could continue procrastinating meal time until it's far too late to do anything other than order out, or you could break the process down into its smallest steps and focus on doing one at a time. To make things even easier, let's say you're thinking about what you want to have for dinner tomorrow, so you have plenty of time to get all the steps done.

The key here is to start with a step so small that you have absolutely no excuse for not doing it. In this example, that would be simply opening up a website that contains recipes or pulling a cookbook off the shelf. This isn't a hard step by any means, but it helps build momentum for the rest of the tasks. Once you have that cookbook or website open, it's second nature to start flipping through looking for something to eat. Find something you like, and then for your next step, look at the directions and decide whether or not you think you can make it. Once you've found a recipe you want to make, write down the ingredients. Then check to see if you have them all, and if not, add what you don't have to your grocery list so you can get them later today or tomorrow. Going to the grocery store is something you likely do every week, so that shouldn't be hard either. Think of this task more as adding one or two extra steps to your usual grocery routine—for example, grabbing some milk or eggs—rather than counting it as the entire grocery trip, which might sound a little more intimidating if you're not especially motivated. Pick up what you need, then bring it home. Then read the recipe through again to make sure you know all the steps. Before you do any cooking, focus on just laying out your ingredients. Then follow each step carefully, and by the time you're done, you should have a completely stress-free meal.

When you focus on each tiny task rather than constantly stressing about all the work that has to go into the finished product, you can sit down and simply get your work done, no matter what it is. Don't worry about what comes next. Concentrate on the present moment, take everything one step at a time, and remember that even a little bit of progress every day is better than no progress at all because you let stress and anxiety get the best of you.

Ticking off Tasks

Many people find that they have an easier time focusing on their tasks and completing them in a timely manner if they make a list and tick things off as they go. The process of making a tick mark or crossing a task out gives you a sense of progression, and you can easily see how much you've done and how much you have left to do. If you use this method, it helps to combine it with the previous advice of breaking a bigger task up into many smaller components, as there's more progress to be made if the steps are smaller. For example, instead of making one task on your list reading 20 pages of a book, you could break this up into five-page chunks. Then you get the little boost in confidence that comes with ticking off tasks four times rather than just once, which can make it easier to get through it without giving up or feeling too stressed out.

Getting Your Stress Under Control

If you don't make an effort to control the amount of stress you're under and how you respond to that stress, it will start controlling you. Maybe you want to attend an event, but you're so stressed about having to socialise with others that you end up turning it down. Maybe you really wanted to finish a project on time for your job, but your stress became too overwhelming and you just couldn't find the motivation to complete it. In many ways, stress can become very restrictive if you don't take steps to combat it. You won't be able to live your life the way you want to if you allow excessive stress to determine what you are and aren't capable of doing.

If you feel like you're being controlled by your stress, there's no need to panic. You can escape from this cycle and start feeling more in control of yourself once again. The first step is to look at your obligations and anything you're dealing with that's causing you stress and see where you can alleviate this. You might have to take a step back from being involved in many things and focus on a smaller set of tasks that you can handle without cracking under the pressure. The next step is to then address how you respond to stress and the resulting anxiety. You are likely never going to be completely stress-free, but if you can handle stress when it appears and channel it into a more positive way of dealing with your problems, you won't end up feeling like you've handed over the reins for your own body. You can maintain control over yourself, calm your anxious feelings, and work through the stress so it no longer holds you back.

Seeking out Sources of Stress

When you feel stressed, you might know that some of that pressure is coming from your responsibilities and other obstacles in your life, but you might not know exactly which ones are causing the most tension. Before you can start alleviating stress, you need to know where it's coming from. Yes, your job keeps you busy, but is it actually stressful, or do you sort of enjoy the fast pace? Yes, you love your family immensely, but is spending so much time worrying about others making it harder for you to relax when you're on your own? Oftentimes, things that you love can still become sources of stress, and things that seem fairly simple can actually become significant contributors to anxiety if there's just too much going on in your life.

One way to cut down on stress is to get into the habit of saying no. You don't have to turn down everything you're asked to do, but you do need to perform an honest review of your schedule, think about what you have time to do, and then avoid overburdening yourself. Taking on too many tasks helps no one, least of all yourself, and certainly not the people who trusted you to put a lot of care and effort into a task that you ended up juggling with five other obligations. When you're overburdened, your work quality suffers because you have to rush through everything and you're distracted with thoughts of everything else you need to do. This isn't a productive work environment, and it's not healthy. Having too many tasks on your plate at a time is a sure-fire way to find yourself completely buried under stress. It's important to recognise that being honest with people who ask for your help and letting them know when you're just too busy is better for both of you.

Learning to say no works for tasks you haven't yet accepted, but what do you do if you've already taken on too much work? In this case, you must learn another important lesson—how to ask for help. Maybe you want to focus hard and try to work through everything on your own, but sometimes there is just too much for any one person to do. Reach out to friends, family, and co-workers and ask if they can lend you some assistance when you need it most; just be prepared to return the favour when the time comes. You can also speak to the people who've asked you to take on these tasks and work with them to reschedule so you can focus on one thing at a time. For example, if your boss asked you to pick up a few jobs and the deadlines end up overlapping, meet with them and ask if you can push some due dates back or have some work reassigned to someone else if you can't handle it. Your mental health is important, and even if you have a particularly stubborn or demanding boss, you need to value your well-being and avoid taking on too much work at one time. Strategically scheduling your tasks so you aren't overwhelmed will help you become a more efficient worker and you won't experience constant anxiety and distractions that would otherwise get in the way of your work.

Calming Yourself Down

While some deadlines can be adjusted so you have fewer sources of stress in your life at one time, others cannot. For example, if you're a parent, you can't exactly take some time off from parenting so you can deal with other things going on in your life. You might be able to call in a babysitter or lean a little more heavily on your spouse for a few days while you catch your breath, but eventually, you will need to come up with a way to manage all of your responsibilities without getting bowled over by stress.

If you're feeling especially anxious or frustrated, it's okay to take a moment to focus on calming yourself down. You might use a breathing exercise, such as a mindful meditation, to reset your brain and quiet the constant jumble of anxious thoughts. Try shifting your focus to the here-and-now rather than worrying about everything you have to do in the future. In some cases, just taking a step back from the stressful situation and doing something relaxing and fun can help you alleviate the pressure you're under. Take a few minutes or even an entire day off, regain your sense of inner peace and balance, and then return to the task with a more positive mindset. Ideally, you should create a routine with a calming activity that you can practice when you get too stressed, whatever that activity may be for you. Take a break, refocus, and then address the issue with a level head.

Taking a Logical Approach

Stress and anxiety often come from overblown assumptions about a problem. If you're having trouble keeping up with your tasks at work, your mind might immediately make the leap to thinking you're going to get fired, you'll be unable to pay your rent or mortgage, and within a few weeks, you'll end up living on the street.

When you're calm, you can tell that this is a less than rational way of looking at an issue of poor performance, especially if you haven't had any prior issues that would lead your boss to fire you. And even if you did end up losing your job, assuming you would become homeless in such a short period of time only seems realistic in the moment because you're jumping to the worst possible conclusions. When you allow anxious thoughts to run rampant in your brain, you become your own worst enemy.

Reframing the problem and potential outcomes in a more logical, reasonable way can help you quell some of your anxieties. Think about what is most likely to happen, not just what you fear is going to happen. Yes, messing up a job could be a problem, but it's much more likely that you might get a light reprimanding rather than getting fired on the spot.

You can extend this logical line of thinking to many other things that give you anxiety. For example, you might be reluctant to do any public speaking because you're worried about how others will perceive you. Maybe everyone in the crowd won't enjoy a speech you have to give, but they're probably not going to boo you off stage just because you seem a little nervous.

You might avoid applying for a job because you're worried the interview will go poorly, but even if it does, the only possible negative outcome is that you don't get the job. If you never apply in the first place, you certainly won't be getting the job, so why not try? Once you start making an effort to predict more rational outcomes instead of defaulting to the worst-case scenario, you'll also internalise the idea that a positive outcome is equally likely. This will inflate your confidence and help you shake off some stress.

Learning how to constructively and reliably bring your stress levels back under control in difficult situations is a key foundation of building mental strength. Learning to tolerate uncertainty, turning big tasks into more bite-sized manageable ones, letting go of the things you can't change to focus on the things you can, eliminating unnecessary sources of stress, and better managing anxious thoughts and feelings are various methods that will help you accomplish the same end goal.

When you practice taking life one step at a time and concentrating on your present reality rather than on the fears and anxieties that live in your head, you'll see that it's not so difficult to reduce your stress levels after all. You can make these strategies even more effective when you combine them with positive habits for becoming more mentally fit.

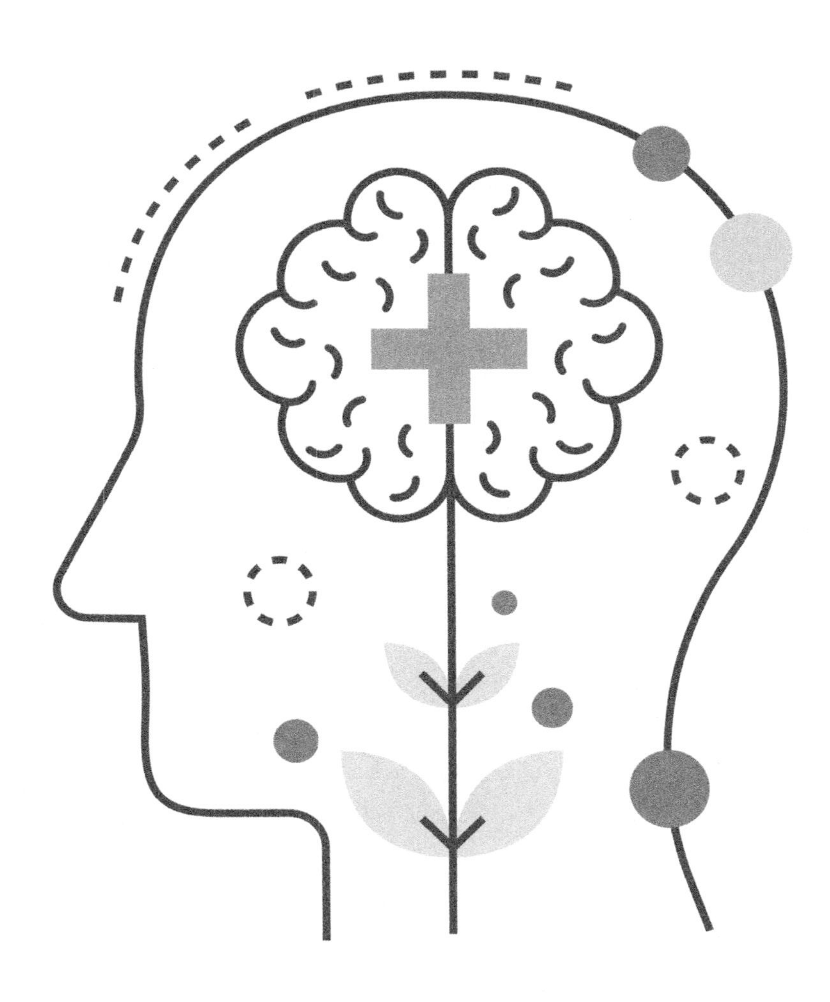

Chapter 6: How to Unlearn Bad Habits and Create New Ones

We often operate under the impression that everything we say and do is a carefully crafted response, unique to the day, but when we zoom out and look at our behaviours on a larger scale, we see this often isn't exactly true. In fact, the majority of what we do each day is actually very similar to what we've done for most of our lives. Think of it this way: when you wake up in the morning, you probably follow the same order of steps you always have, and this doesn't change unless there's something preventing you from practicing your usual morning routine. If you're a coffee or tea drinker, you likely have at least one cup before you leave for the day without fail, or you routinely stop for a drink on your way to work. You probably shower, get dressed, brush your teeth, and prepare breakfast and possibly lunch in the same order. You might even tend to grab the same thing for breakfast each day, or skip it entirely on a regular basis. In short, these routines are the foundation of your behaviours. If you don't have any reason to change them, you probably won't.

These repeated behaviours are known as habits, and they influence a significant portion of how we think and act each day. According to a 2014 study conducted by the Society for Personality and Social Psychology, "about 40 percent of people's daily activities are performed each day in almost the same situations," (Society for Personality and Social Psychology, 2014, para. 1) but some estimate that as much as 95 percent of our behaviours may be influenced by habits in some way. This can be used as a positive tool if your habits help you remain healthy, happy, and mentally fit. However, habits can also play a negative role in your life if you don't realise how much power they have over you. Failing to properly examine your habits allows bad ones to fester, gradually leading you down the path of mental health issues and often physical health issues as well.

To really take control of your mindset, you need to learn how to identify the bad habits you're currently practicing and recognise what makes them so harmful. Then, once you know where the problems are, you can begin the process of walking back these habits and replacing them with new and improved ones. These positive, healthy habits will help you automatically practice the behaviours that will improve your mental strength and ditch negative acting and thinking patterns.

What Separates a Bad Habit From a Good One?

Bad habits that go completely unrecognised are often the most dangerous of all. If you don't learn how to spot and acknowledge the difference between bad habits and good ones, you won't be able to figure out which behaviours you need to change and which you should reinforce. Of course, it is not always as black and white as you might think, especially when talking about habits that are bad for your mental health. If you smoke, it's very obvious that this is a bad habit. You might prefer to give up smoking for the sake of your health, but if you're engaging in harmful coping mechanisms, you may not even recognise the damage these actions are doing.

To understand what makes a habit good or bad, you must consider what the habit helps you achieve and what its unintended side effects are. Is it a net positive in your life, or does it represent a temporary solution that prevents you from actually dealing with a bigger problem? For example, you might find yourself constantly seeking reassurance from friends and family before you have to make a big decision. While it's okay to ask for opinions every once in a while, if you do this so often, it's second-nature for you to ask others when you need to make a choice and you're actually undermining your own ability to make decisions. You might subconsciously value your opinion less than others' thoughts and feelings, so you never give yourself the chance to make any choices, good or bad. In this way, this seemingly-innocuous habit can actually be quite harmful as it prevents you from achieving the goal of being more self-confident and independent.

Progress Towards Goals

Many habits are negative influences on your life because they impede your progress towards your personal and professional goals. A good habit helps you achieve what you want out of life, while a bad habit actively impedes your progress, whether you recognise its detrimental effects or not. When evaluating whether a given habit is helpful or harmful, it's necessary to look at its cumulative effect if you practice it for many months or even years, not just the effect it has in the moment.

Every habit starts out as a single action that seems fairly harmless at first. For example, a habit of drinking multiple cans of soda each day starts with having a single glass. Habits get their power from repetition, though, which means you need to look at the cumulative effect. One glass of soda may very well be nothing to worry about, but if you have one each day, this can start to cause problems like high blood sugar, a higher risk of developing diabetes, and a caffeine addiction, which only fuels your need to engage in the bad habit. A single soda may not prevent you from achieving your health goals, but turning it into a habit will. Therefore, it's necessary to consider how a habit's negative effects build up over time and what kind of damage it's causing, whether the toll is physical, mental, or emotional.

Think of all the goals you want to achieve and how you're inadvertently holding yourself back from them. Just by picking up this book, you've proven that you have a desire to encourage positive change in your life. If not, you wouldn't be concerned about your mental well-being. You likely had a few objectives in mind when you decided you wanted to become more mentally fit. Maybe you want to use mental health as a vehicle for improving your overall health. Maybe you want to have a more positive mindset, or to routinely practice kindness towards yourself and others. Maybe you desire confidence and personal success, or maybe you're looking to use these traits as ways to get ahead in the professional world. Whatever your goals are, you need to recognise how engaging in poor habits are holding you back from achieving them. Consider how regret, guilt, and shame build up and prevent you from trying new things or being compassionate to yourself. Notice how every time you back out of a project or second-guess yourself, you're allowing your negative thoughts to win. As you start to identify the bad habits in your life that contribute to poor mental health, you'll also see how these habits serve to prevent you from achieving the goals that motivated you to start your journey of self-improvement, and why it's so important to quit them once and for all.

Understanding Why Bad Habits Occur

Paying greater attention to the effects of bad habits is just one piece of the puzzle, of course. It's also necessary to understand the cause of these habits and how they've come to have so much power over you, so you can work to take that power back. Bad habits can form because of a variety of different reasons, but they are often related to a need in your life that you're trying to fulfil the wrong way. For example, you might feel the need to be praised by others because your self-esteem is low, so you fall into the trap of endlessly scrolling through social media and trying to curate the perfect feed, even to the detriment of the other responsibilities you should be more worried about. Getting the notification that someone liked or commented on your post gives you a little boost of dopamine, a brain chemical associated with happiness and learning. This trains your brain to engage in the habit again and again because you get a tiny spike of happiness and validation from it. But at the end of the day, these spikes are temporary. If you pin all your hopes of feeling validated and loved on the interactions you get from social media, you'll fail to look inward for a more renewable sense of strength, and you'll always be dependent on someone else to provide what you should be giving to yourself. In short, bad habits frequently create positive feelings, but they don't change the problem that created your negative mindset in the first place, and therefore, they only offer temporary relief. It is only by breaking your bad habits and replacing them with better ones that you can start to address the underlying issues behind feelings of dissatisfaction and unhappiness. This allows you to practice more long-term strategies for lasting happiness and success in your life.

How to Quit Bad Habits

So, bad habits are bad news, and chances are you've just recognised that you have quite a few harmful habits you didn't realise you were engaging in before. What can you do to quit them? According to behavioural science research, it is incredibly difficult if not impossible to simply quit a habit. This makes sense, as habits are, by definition, actions you've taken for so long that they have become automatic responses to given situations. If you encounter these situations again, chances are you're going to revert back to your old ways if you don't have an alternate plan. Instead of quitting bad habits, you should focus on replacing them with better ones.

Theories about behavioural change point to the importance of 'overriding' bad behaviours with better ones, as this is much easier than trying to break down the connection in your brain between any problem and the bad habit spawned from it. In fact, it may even be impossible to truly unlearn any habit. According to research reported in 2012 by the U.S. National Institutes of Health, "replacing a first-learned habit with a new one doesn't erase the original behaviour. Rather, both remain in your brain. But you can take steps to strengthen the new one and suppress the original one" (Wein & Hicklin, 2012, para. 17). The best way to do so is through repeated practice, which strengthens the connections between a given habit trigger and the new, positive habit in your brain until you start to default to that one instead. But to understand how this works, you must first understand the habit loop.

The Habit Loop

You now know why bad habits take hold and why they're so dangerous. The next step in walking back these bad habits is understanding how they perpetuate themselves so you know how to interrupt the cycle. Habits are a product of your situation or current environment. You start to associate your habits with different cues, known as triggers, and these become so closely entangled that you begin automatically performing the habit whenever you encounter the cue. For example, when you wake up in the morning and have morning breath, this is your cue to brush your teeth. Bad habits can have cues as well. Maybe you picked up smoking to help you deal with anxiety, and now every time you feel anxious, you also tend to pick up a cigarette. Maybe instead you started smoking in a social setting, so any time you are outside with a small group of people, you feel the urge. The habit cue sets off your desire to actually perform the habit, and the more you strengthen the mental connection between that cue and a specific behaviour, the harder it becomes to break.

The next stage of the habit loop is the routine. This is the action you take as a result of the cue. In other words, it is what you perceive as the habit itself. In the previous example of outdoor social gatherings functioning as a cue, the routine would be smoking a cigarette. The routine is the part you'll need to change, but you can't do so without understanding the cue that spurs it. Oftentimes, the routine is a way to address a problem or alleviate some deficiency caused by the cue. For example, maybe your cue to habitually drink coffee is when you wake up in the morning after getting poor sleep. The coffee is addressing the issue of you feeling exhausted. If you didn't feel quite so tired, you might not feel the pressing need to reach for caffeine first thing in the morning. If you can't change the cue, however, you must alter your response to it—the routine—or you will inadvertently reward yourself for bad behaviours.

The final part of the habit loop is the reward. This is the results of the habit that encourage you to continue performing the habit every time you encounter the same cue. When you're feeling low, you might reach for candy because it tastes sweet and it gives you a little dopamine boost, making you feel better. This is just a temporary fix and can be harmful in the long run if it's a habit, but your brain learns to anticipate the reward of getting your sugar fix, which becomes more enticing than any far-off future benefits of refusing the sugar. Every habit has a reward, even bad ones. These rewards perpetuate the habit loop and compel you to repeat the same cue recognition and corresponding behaviour again and again. It is typically hard or sometimes impossible to divorce a reward from a given habit because you generally don't have control over them unless you are creating your own rewards. Therefore, it is usually more effective to focus on the cue and what behaviour you take in response to it if you're looking to replace an old habit with a new one. Everything starts with identifying the conditions that encourage you to practice your bad habits.

Identify the Habit Trigger

A trigger is the motivation behind any habit, good or bad. These can come in many different forms depending on the sort of habit you're performing. A tendency to check your phone repeatedly throughout the day instead of working might be a result of boredom, or you might have a tendency to hunch in on yourself whenever you're in a social situation that's causing you a lot of anxiety. Triggers may be feelings, other peoples' actions, your own actions, things you notice in your environment, times of day, locations, states of mind, or anything else that encourages you to act or even think a certain way. Something as small as seeing a box of biscuits on the counter could encourage you to indulge in them, while other people may not reach for the junk food unless they're experiencing stress and frustration. Pay attention to your unique habit trigger so you can anticipate the practice of a bad habit and eventually start replacing it with a better one.

Identifying the habit trigger comes with a secondary benefit, which is a clue as to what kind of good habit might make a good replacement. Your habit trigger tells you what you're trying to accomplish by performing the habit. A trigger of stress means you're trying to help yourself relax without taking a full break, which means it may be beneficial to replace the behaviour with one that helps you destress in a healthier way. A trigger you only perform in the evening, like having a nightcap before bed, might indicate you have difficulty getting enough rest and you need to develop a less self-destructive night-time routine. The better you understand the motivation behind any behaviour, the easier it becomes to leave that behaviour behind in favour of one that addresses the issue without harming your health and mental wellness.

If you're having difficulty figuring out just what trigger is resulting in negative behaviours, you can try keeping a habit journal. Write down every time you engage in a habit you're trying to break, as well as when it happened and how you were feeling at the time. Make a note of anything else unusual that might have been going on, including any especially loud thoughts. Over time, you should start to notice a pattern emerging. You might repeat the habit around the same time every day, or maybe only when you're experiencing a certain emotion. The habit might serve to improve your mood or energy levels, in this case. You might find that you only engage in the habit on days when you have to go to work, or when you have to see a certain person. These can indicate that stress might be a component in your habit trigger. Whatever the habit cue is, this method should help you pinpoint it by encouraging you to be a little more mindful of your thoughts and actions, thus no longer allowing the habit to be completely automatic.

Reflect on the Harm Caused by the Habit

A common mistake many people make when trying to change their habits is failing to reflect on why they really desire the change. Yes, you want to change a habit typically seen as 'bad' and replace it with something better, but are you doing this because someone told you to, or because you really want to improve? External motivation will only take you so far. If you want to create lasting change, you need to be internally motivated, and this means working to recognise why your bad habit is so detrimental, even if it doesn't seem so bad at first.

Consider a habit like multitasking. If you want to be productive, you probably started using multitasking as a way to improve how much you could get done in a short amount of time. A boss, parent, or someone else in your life might have even told you that multitasking would help make you more efficient, so whenever you experienced the trigger of getting a little overwhelmed, you would look to multitasking as the answer to your prayers. Unfortunately, science doesn't quite back this up. Rather than helping you accomplish more, multitasking might actually be interfering with your ability to focus on any one task and ultimately slowing you down as a result. This can add even more stress to your plate and further overwhelm you. However, if you don't recognize how destructive this bad habit can be, you may not feel very motivated to overwrite it. You might even assume multitasking is helping you, when this is likely not the case. If you find you need some extra convincing, try to perform a little experiment with your habit to see whether it's really helping or hurting you. For multitasking, this might mean trying to make progress on the same set of tasks while multitasking and while taking them one at a time, and testing to see which method is faster. For other habits that are more connected to your emotions, you might track your mood throughout the day so you can see how indulging in bad habits only provides a temporary mood boost, and how you might even feel worse later on.

Ultimately, you are not going to make any changes unless you feel they are necessary, especially when the change is as difficult as replacing a habit. Therefore, you must demonstrate the necessity to yourself. This will ensure you have all the motivation and willpower you need to overcome temptation in the future while you work to improve your daily habits.

Replace the Habit With a Better One

The final step in interrupting a bad habit loop is to swap out the habit in question for a better one. If you want a habit that will really stick, you'll want to choose a behaviour that also follows naturally from the same cue. For example, if the trigger for the bad habit is feeling stressed, your new habit should be something else that helps you alleviate tension, like breathing while counting to 10 or performing another restorative activity like a creative hobby. If boredom leads you to reach for your phone at work, try to do something more engaging with positive benefits like getting up from your desk and doing a few jumping jacks, or switching tasks to something a little less draining. Over time, the connections between this new habit and the old cue will start to strengthen, but in the meantime, you want to make it as easy to shift gears as possible. By picking a new habit that is strongly associated with a given trigger, you don't have to relearn an entirely new habit loop. You just have to swap out the 'response' part of the loop for a different habit.

This process is a little different if you want to build a good habit out of nothing. For example, your bad 'habit' might be not exercising. While you can consider a lack of motivation to be a trigger for engaging in this practice in a way, it's not a cue in the traditional sense, so it's difficult to swap out inactivity for exercise without giving yourself something to work off. In this case, you might benefit from adding a clear cue that leads you to start working out. You might decide that you're going to exercise first thing in the morning so you can start your day with as much energy as possible, which means waking up would become your cue. Reinforce this by always exercising at the same time and refraining from skipping a day until the habit is firmly in place. Alternatively, you might create a habit trigger by changing part of your environment, like leaving running shoes near the door so you notice them as soon as you get home from work. Upon seeing the shoes, you'll be reminded of your commitment to exercise after work, which cues the habit.

In either case, replacing a habit means moving away from an old, undesirable way of life and embracing a new and exciting one. It means improving your life, even if you have to start with baby steps and gradually build up the positive habit over time. When you really focus on your habits, identifying and removing bad ones in favour of ones that help you stay healthy and happy, the benefits will be exponential, all because you chose to embrace good habits.

Building up Good Habits

Oftentimes, negative habits form without your express consent. You didn't mean to encourage your sugar addiction, but by having cakes or biscuits around in the house and indulging frequently, you accidentally reinforced the negative behaviour. You didn't intend to respond to stress with thoughts or actions that only made your fear and frustration worse, and yet that might be exactly what you've done without realising. This means that when it's time to replace these bad habits with good ones, you might not have a clear idea of where to begin. This is especially true if you want to achieve a certain mental health goal and you're not sure how to get there.

If you devise the right habits, you can slowly and steadily work towards your goals. Every time you practice a good habit, you get a little more well-versed in it, and the good behaviour starts to become second nature to you just as a bad habit might have become second nature before. As you continue to practice positive mental health habits, you can learn how to utilise healthy coping mechanisms to your advantage to defeat fear, anxiety, worry, and insecurity. It all starts with learning how to take advantage of the natural pathways in your brain.

Brain Plasticity

Your brain is a highly adaptable organ. It is capable of taking in new information and completely changing the way you react to given situations. It's also capable of learning and retaining new information, often through repetition. This is known as plasticity, which describes your brain's ability to adapt to new experiences so you can better handle them in the future.

Plasticity plays a significant role in habit development. This is because it relies heavily on repeating information or actions and changing your behaviours based on feedback. You've likely noticed that when you hear the same thing multiple times, you're more likely to remember it. You might not have a clear memory of a song the first time you hear it, but play it back a few times, and you'll likely know all the lyrics before long. This is doubly true if you can connect your newfound knowledge to a specific experience. You might remember listening to the song in the car with a friend or parent, maybe belting out the words, so it's a lot easier to remember them. This is similar to how a certain sensation like smell can almost take you back in time to past events—for example, the smell of fresh-baked biscuits might remind you of the ones your mum used to make almost instantly. This is just a by-product of the way your brain solidifies memories: through repetition and association. You can use your brain's plasticity to your advantage when trying to get new habits to stick.

The core foundation of a habit is repeating it constantly in response to the same cue. The cue is the 'association' part of plasticity. You associate a given thought or action with the cue or trigger, and it causes you to perform the behaviour, as long as you train yourself to do so first. The training comes through repetition of the habit. If you can consistently perform the same behaviour over and over again, you'll have an easier time getting it to stick. The more you use the same neuron pathways in your brain, the less trouble you'll have recalling information or reminding yourself to do something. Through repetition, you're exercising the neuron pathway connected to the behaviour over and over again. This practice is what makes habits stick, and it's what you'll need to use to your advantage if you want to create and maintain good habits.

Repetition and cue association are incredibly helpful aspects of habit building to know, but they're just the basics of the habit-forming process. To really get your habits to stick, you'll be best served by starting small and working your way up to bigger habits over time.

Start Small

Consider this scenario: you want to make an effort to get healthier, so you decide that you're going to make it your New Year's resolution. On the first day of January, you make yourself a salad for lunch and a nice healthy meal of grilled chicken for dinner, and you do a very high-intensity workout for an hour. At the end of the day, you're exhausted and kind of hungry, but you stuck to your plan. The next day, you're sore and sort of craving the less healthy lunches you used to eat already, but you commit to eating healthy again. This time, you can't quite make it through your whole exercise, so you do about 45 minutes before getting tired and deciding you've done enough. A week later, and the thought of having another salad makes you feel vaguely ill. You're so tired and sore that exercise sounds impossible too. You decide you deserve a cheat day, so you binge on some junk food and lounge about the house, skipping your workout. You have every intention of picking the healthy habits back up, but you haven't practiced them for long enough to make them stick, and you just can't work up the motivation to push yourself that hard again. Less than a month after getting the idea to embrace healthier habits, you've given up on them.

What went wrong here? The main issue is that you chose to start many habits all at the same time, and many difficult habits at that. Instead of easing yourself into things, you dropped yourself in the deep end, and you quickly found yourself drowning. Keeping up with many new habits is very overwhelming, and it doesn't give you the time and space to properly adjust to any one habit. This also means you can't efficiently associate any one habit with a given cue, so you're not using the habit loop to your advantage. You will quickly confuse and overwhelm yourself, and if you tire yourself out like this, you'll give up on the habits before you've made any progress. You need to give your habits time to build, like working a muscle. Start with small habits so you can first focus on repeating them every day alongside their appropriate cue. Once you have the cue and mini-habit down, you can gradually increase the scale and intensity of the habit until you start to see positive results.

Let's look at the same example from this improved perspective. Rather than trying to implement healthy eating and exercise at the same time, choose just one habit to focus your efforts on for now. You can add the other healthy habit to your schedule once you have the first one down. Let's say you decide to start with creating an exercise routine. Instead of planning out a very intense hour-long workout that leaves you sore and cranky with little motivation, start with something so small you can do it without trouble. Maybe begin with five minutes of jumping jacks, or if you want something less focused on cardio, you can try lifting five-pound dumbbells or doing a few crunches. Create a cue for your workout, such as starting it at the same time each day. Do your five-minute workout, then come back the next day and do it again. Repeat this process, being as consistent as possible, until you feel like you have the hang of it. Already, exercise should start to come a little more naturally to you since you're used to doing it on a regular basis. After a week or two, you can increase the intensity and the length of your workout. Maybe do a 10-minute workout for another week, then move up to 15 or 20 minutes, and so on. With each increase, you'll be building up both muscle and willpower, which will help you actually stick to the habit even when you feel tired and unmotivated. This will solidify it as something you do every day. Once you feel like you've mastered the new habit and reached the level of performance you want to be at, you can move onto another healthy habit, repeating the process all over again.

When you start small, you're less likely to tire out. You'll be able to rely on your willpower rather than depleting it entirely in a short amount of time, which makes a huge difference in how often you practice the habit. After maintaining a habit for a few weeks or months, the repetition helps the habit stick in your mind, which goes a long way towards preventing yourself from giving up and falling back into bad habits.

Avoiding Backsliding

In any effort to improve yourself, there will be moments where you don't quite meet the mark. In habit formation, this means slipping up and returning to bad habits again. To weaken a habit, you need to avoid practicing it, so returning to your previous habits can make it especially difficult to actually quit them. Therefore, you'll want to take steps to prevent yourself from backsliding as best as you can. This often means reducing the amount of temptation in your life, including the presence of stress.

Excessive amounts of stress often give temptations more power. If you tend to practice bad habits to alleviate negative emotions, being under a lot of pressure is going to make those negative emotions more prevalent and harder to ignore. As always, it's important to take some time to relax and find more positive, constructive ways of coping with difficult moments in your life. Rely on friends and family when your own willpower runs out. Communicate about your problems and ask them to avoid inadvertently tempting you or placing too much stress on your shoulders, and allow them to help if they offer their support. Find time each day to sit and reflect, coming to terms with your emotions and developing a better understanding of yourself so stress won't drive you back into your bad habits. The better you reinforce your positive habits and look for healthy coping mechanisms, the less stress and temptation will be able to lead you astray.

Ultimately, the most important thing to remember about backsliding is to not be too hard on yourself when it occurs. We all make mistakes sometimes, and while you should do your best to avoid engaging in bad habits again once you've made the decision to drop them, life doesn't always go according to your plans. There may be times where temptation becomes too strong and you give in, or when you're feeling too low to even consider following healthy habits. If you harshly blame yourself and call these mistakes failures, you'll never want to try again. It's more important to show yourself compassion and understanding, and to recognise that at the end of the day you are human and you may not always do the best things 100 percent of the time. By showing yourself compassion, you can accept your flaws, recover from your missteps, and set yourself back on the path to positive progress.

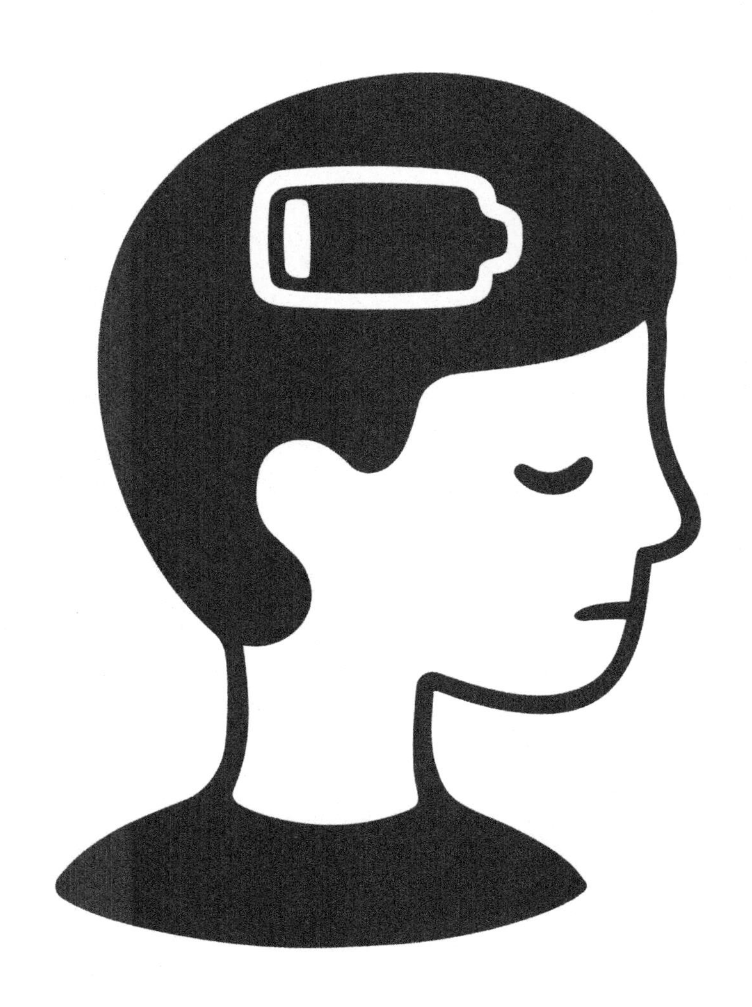

Chapter 7: How to Sleep at Your Best

Poor mental health and poor sleep are more connected than you might initially assume. If you have trouble sleeping, you might notice that you can't seem to be able to get your brain to turn off at night. Maybe you're constantly thinking about some upcoming event that has you nervous, or your thoughts keep circling around all the work you have to do. Alternatively, your thoughts might turn to the past. You're all ready to drift off, when suddenly your brain sees fit to remind you of something embarrassing you did last week, or even a regretful event that happened many years ago. When you start fixating on these thoughts, fuelled by anxiety, worry, and rumination, it becomes incredibly difficult to actually find the peace of mind needed to fall asleep. You might find yourself lying in bed staring at the clock as the numbers continue to tick up, thinking about all the sleep you're missing out on because you can't manage to relax.

Not getting enough sleep or having poor quality sleep can, in turn, make it more difficult to support your mental well-being. You might wake up "on the wrong side of the bed," frustrated and grouchy because you haven't gotten the rest you need to operate at full capacity. Little annoyances might get under your skin more than they would normally, and you may have a harder time letting go of any negative thoughts. You'll likely feel more stressed as you go about your day. Your work quality may suffer because you have trouble focusing on your tasks. This can feed into negative thoughts about yourself, contributing to feelings of shame and self-hatred that will only worsen conditions like anxiety and depression. In short, mental health issues can contribute to sleep issues, and sleep issues can contribute to mental health issues. This creates a cycle that's very difficult to disrupt, but not impossible.

Luckily, if poor mental health can have a negative effect on sleep, then this means taking care of your mental well-being can improve your ability to go to bed and feel well rested in the morning. Similarly, developing better sleeping habits is a significant contributing factor to getting your mental health issues under control. With better rest, you'll feel more equipped to handle difficult and stressful situations, and you'll have less trouble concentrating on the tasks that need your attention throughout each day. By focusing on your mental state and night-time habits, you can establish a sleeping routine that is more conducive to mental fitness, ensuring you're as well-rested as possible so you can face the challenges of every day head-on.

Entering Deep Sleep

A great deal of sleeping advice starts and ends at "get at least eight hours." This is a good precedent to set, but it's often not enough to ensure your sleep is truly restorative. Have you ever gotten eight or more hours of rest only to wake up just as tired as you were the night before? You might be getting enough sleep, but if the quality of that sleep is very poor, it's not going to do much for your physical and mental health. To really benefit from the sleep you're getting, you need to start sleeping deeply.

Why Deep Sleep Matters

Not all rest is created equal. You might have noticed that it's very easy for someone to wake you up if you've just drifted off, but it can be harder to force yourself awake if you've been sleeping for a few hours. This is because of the different sleep stages your body and brain experience while you're dosing. These are divided into the three non-rapid eye movement (NREM) stages and a rapid-eye movement (REM) stage. NREM stage 1 occurs as you're just drifting off. Then you progress into NREM stage 2, where your heart and breathing rate begin to even out. NREM stage 3 involves the deepest kind of sleep, when your muscles relax and your brain becomes less active. After this, you move back into NREM stage 2, before finally progressing to REM sleep, when brain activity picks up, which leads to dreaming.

It is critically important to reach NREM stage 3, as this is when your body and mind undergo the most healing. It is the "sleep stage responsible for healing and repairing your body, replenishing cells, and revitalizing your immune system" (SleepScoreLabs, 2020, para. 1). If you consistently sleep in short bursts, waking up at the slightest noise and having trouble falling back asleep, you won't feel alert and energized when you wake up. Additionally, if you tend to go to bed late and wake up early, depriving yourself of sufficient amounts of sleep, you won't give your body the chance to remain in this stage to fully rest. To make sure you're getting all the deep sleep you need to heal your mind and body, you need to establish a night-time routine.

Creating and Sticking to a Bedtime Ritual

Chances are you haven't thought about holding yourself to a "bedtime routine" since you were a toddler. However, without a proper period before bed where you wind down from the stresses of the day and prepare yourself for sleep, you'll likely find it very difficult to calm your thoughts. This may be a large contributor to lying in bed without being able to get any rest. While your body may be inactive, your mind certainly isn't because you're still thinking about everything you were doing right before bed and everything you need to do in the morning. This is never going to result in high-quality sleep, as even if you do drift off, your stress and anxiety might find their way into your dreams and turn them into worry-fuelled nightmares.

It might feel a little silly at first, but setting aside some time at least a half hour before you want to go to bed to relax, unwind, and get into the right mindset for sleep can do wonders for encouraging a deeper rest. This process primarily involves setting aside stressful and distracting things, performing a relaxing activity, and making sure the conditions of your bedroom are ideal for promoting deep sleep.

Setting the Stage for Good Quality Sleep

Some people can fall asleep just about anywhere, while others find it practically impossible to go to bed unless they fulfil a set of incredibly specific requirements. No matter which kind of sleeper you are, it's true that there are some conditions that are more optimal for promoting high-quality rest than others. If you're having trouble feeling well-rested throughout the day, it's a good idea to start by replicating these ideal conditions as best as you can.

First and foremost, remove as many distractions from your sleeping environment as possible. This includes the TV, as late-night programming often includes sudden loud noises and bright lights that can jolt you awake when you're just starting to drift off. If you need some background noise, you can try replacing TV with a white noise machine or some calming videos specifically designed for promoting rest, such as those available on dedicated apps that host relaxing stories and autonomous sensory meridian response (ASMR) content. While it's fine to use your phone to play these kinds of videos, you should avoid actually looking at your phone screen to surf the web or play games for at least half an hour before bed. Getting caught up in games and social media feeds can cause you to procrastinate getting some rest, but more than that, the blue light emitted by phone and computer screens could potentially disrupt your sleeping patterns. Some devices include blue light filter options, but it's best to forego screens altogether before bedtime.

You should also ensure you're creating an ideal sleep environment for yourself. This means your room shouldn't be too hot or too cold, and it shouldn't be too brightly lit either. If you sleep during the day because you work at night, you can replicate an appropriately dark environment with blackout curtains. You'll have an easier time falling and staying asleep if you're in a quiet, cool, and dark room where you feel completely safe. This is why it's such a great idea to turn your bedroom into a zone of rest and relaxation, as mentioned in Chapter 4. Avoid doing work or other tasks that require you to focus on your bed so you only get into bed when you're ready to sleep. This will help strengthen the connection between your environment and getting a good night's rest so you can fall asleep quickly when you finally decide to lay down.

Quelling Racing Thoughts

This scenario probably sounds familiar: you have your environment carefully crafted to meet the optimal sleeping conditions, you're in bed, but you just can't get your brain to turn off. Rumination and other racing thoughts won't seem to go away. Why does this happen, and what can you do to let go of these thoughts and finally get some rest?

Night-time can provide the perfect opportunity for self-reflection because all of the distractions from the day have been turned off and set aside. You're not keeping yourself busy, so your thoughts turn inward.

This is a perfectly natural process, but it can become disruptive if you struggle with mental health issues because your thoughts might be less than helpful. You might fixate on regrets and fears, or start worrying about something with a complete inability to let these concerns go. Worse, you might not be able to focus on a single thought, like you would when performing a meditation or breathing exercise, as exhaustion can interfere with your ability to concentrate. Your mind bounces from negative thought to negative thought, never really letting you process any of them so you can set these anxieties aside. As a result, you have difficulties falling asleep when your brain gets caught up in these worries. Even if you do eventually drift off, fears and anxiety can invade your dreams and lead to tossing and turning or sleeping in fits and starts, which is hardly conducive for high-quality sleep.

There are a few different ways you can ease these racing thoughts, such as trying out a progressive muscle relaxation exercise or listening to a guided sleep meditation. One of the most effective methods is creating and following a specific series of steps that you perform before bed to put you in the right mindset for sleep. Think of the way you might get ready in the morning. You wake up, stretch, brush your teeth, take a shower, get dressed, eat some breakfast, and finally drive to work. Each of these steps helps mentally prepare you for the idea of working so that by the time you get to your job, you're in the right headspace. You might still be a little drowsy, but you're much more awake than you would be if you hit snooze on your alarm a few times and refused to get out of bed. This is why you might feel more tired on weekends, even if you get a few extra hours of sleep.

Now, use this morning ritual as an outline for creating a night-time ritual. If you go right from watching a movie or working to lying in bed, chances are your brain will struggle to make the switch right away. This means you'll be unable to turn your brain off, since your mind isn't really ready for sleep. You can prepare yourself by creating a routine that culminates in lying in bed. Start by getting into your pyjamas an hour or so before bed. Brush your teeth and take care of any other hygiene needs. Set your phone or laptop aside. Switch over to a calming activity like reading to slow your heart rate and lower your blood pressure. Then, when you feel like you're starting to drift off, get under the covers and focus on deep breathing. Over time, your body will learn to recognise these habits as predecessors to getting into bed, so by the time your head hits the pillow, you should already feel pretty sleepy. With this method, you don't give yourself the chance to fixate on your worries because you're altering your mindset. All your thoughts are focused on sleep by the end of the routine, and there's no time to indulge in stressful imaginings.

Sleeping at the Same Time Every Night

Higher quality sleep can be achieved if you start going to bed and getting up at the same time every day. Even if you can get eight hours each night, those hours might not be quite as restful if you lay down at completely different times. Having a set bedtime and making an effort to stick to it makes it easier to fall asleep right away, as you've trained your body to get drowsy around the same time each night. This means less time spent tossing and turning trying to find the perfect position to finally drift off. Creating more positive habits in general is a great way to boost your mood and improve your ability to get more done in a day, and sticking to a good sleeping routine is just another habit that contributes to this overall wellness.

If you can't fall asleep, get up and walk around for a bit, and don't return to bed until you feel appropriately tired. This may seem counterintuitive, but even though you want to get to bed on time each night, it does you no good to lay there for endless hours or to resort to scrolling on your phone when you can't quiet your thoughts. It's better to leave the bedroom for about 30 minutes, do something calming that doesn't involve screen time, wait until you begin to feel drowsy, and go back to bed.

If you end up falling asleep a little late, avoid the temptation to sleep in. You might feel like you need the extra sleep, but you're actually sabotaging your ability to go to bed on time the following night. If you keep waking up later and later, you'll keep falling asleep later and later too, and this sets a bad precedent that will eventually ruin your carefully crafted sleeping schedule. Set an alarm and avoid sleeping in for more than an extra hour or so on weekends. If you start really improving your sleep quality by sticking to a regular schedule, you shouldn't need the extra sleep anyway.

Sleep may seem like simply something you have to do that wastes a good chunk of your day, but it's important to remember all the benefits sleeping provides. Getting a good rest cannot be substituted for caffeine if you want to feel your best each day. When you start prioritizing high-quality sleep, you'll find it's easier to manage your moods, and you can embrace each day with plenty of energy and a positive attitude.

Chapter 8: How to Use Positive Mental Attitude to Help Manage Your Physical Health

We have touched on the value of maintaining your physical health alongside your mental health throughout the entirety of this book, but the importance of staying physically fit cannot be overstated. Your mind and body are two pieces of a single unit, and one cannot function properly without the other. If you're struggling to improve your physical health, you might not see positive results until you really work on improving your mental health. Similarly, if you start neglecting your physical health, even if you think you're doing yourself a favour by "taking a break" and only focusing on your mental well-being, you're actually undermining your mental and emotional strength. True mental fitness means caring about your body alongside your mind. Luckily, all the steps we've discussed so far to help cultivate a positive mental attitude will assist you in sticking to healthy routines for your physical self as well.

Everything in your life is affected by your mindset. You can turn an incredible day into one full of annoyances and frustrations just by focusing on the wrong things. You might have a fun-filled day ahead of you, but a little overcast weather or an aggravating conversation can completely sour your mood if you're going through life with a negative mindset. If you want to find a fault in even the most perfect moment, you can certainly find one, but why spoil your mood? Sure, having a more pessimistic outlook means you rarely give yourself the opportunity to be disappointed because you never get your hopes up in the first place, but this is typically a terrible way to live, as you miss out on so much that you would have otherwise enjoyed. If you commit to having a negative mindset, you're going to look back at your life and view it in terms of only the worst moments, which makes it incredibly difficult to find the motivation to improve yourself. If you assume you'll never get better, mentally or physically, you're not going to put in the work to improve yourself, creating a self-fulfilling prophecy.

On the other hand, having a positive mindset can help you see the silver lining that comes with every cloud. Even when bad things happen, they won't have enough power over you to stop you for long. With a positive mindset, you'll find it easier to stick to positive habits even when you're not having the best day, simply because you're so excited about all the good work you're doing. You'll start looking forward to challenges rather than dreading them because they give you an opportunity to prove yourself and expand your skills. You'll come up with a solution for every roadblock you encounter, all because you've chosen to look on the bright side and never let the day-to-day doldrums of life get you down for long. Of course, a positive mindset doesn't mean you can never feel sad, angry, or worried, but it does mean that these feelings won't become so overwhelming they ruin your day. You can accept them, experience them, and then let them fade, knowing there will always be better days up ahead.

A positive mindset will see you through countless challenges when you're working on improving your physical health. You'll have a significantly easier time eating, sleeping, and exercising right, as you'll continually recognise how much progress you've already made and how much better you feel as a result. You won't struggle to find the willpower to get up out of bed and get active; instead, you'll find that it's readily available. All you have to do is learn how to harness it to take control of your dietary and exercise habits so you can build a healthier, happier life for yourself.

Harnessing Your Willpower

Willpower is one of your most valuable resources. It will help you find your way through the darkest of tunnels to embrace the light on the other side. It will also help you manage regular, everyday problems that might stand in the way of your ability to improve your physical fitness.

One of the biggest reasons people give up on their physical health is because they lack the willpower to commit to good habits. They try these habits out for a few days, but when it gets more difficult to maintain them, they run out of their original burst of motivation and just give up. Think about the example we covered in Chapter 6 where it was easy to give up on prioritising nutrition and exercise when they became too hard. Without a solid habit foundation, when the going gets tough, you will simply give up.

You want to give yourself all the tools possible to avoid this unfortunate outcome. This means drawing strength from within and really caring about the good work you're doing. Set goals you're genuinely passionate about, even ones that surpass simple physical fitness, and create milestones to help you track your progress towards them. Avoiding giving up also means removing the obstacles standing between you and success. Learn how to harness your willpower to ignore temptations or remove them from your surroundings to make it even easier on yourself. Finally, if your willpower ever does run low, don't be afraid to rely on others for help when you need it. Following these steps will help you make the most of willpower so you can always maintain your focus on your goals.

Ignoring Temptations

When you're focused on your physical health, the temptation to take a "cheat day" can be incredibly difficult to ignore. This gets even trickier if you're under a lot of stress. You might not have the patience to cook a whole healthy meal, and you would much rather order unhealthy takeaways instead. You consider working out, but you decide you would be better off kicking back and watching TV. You pass by that bowl of candy so many times you can't help but take a piece.

These are all examples of caving to temptation, and they can make it seriously difficult to actually achieve your health and fitness goals. Temptations are hard to ignore because they play to your weaknesses. If you're trying to develop healthier habits where you eat fewer sugary foods and work out more often, then sugar and foregoing exercise are going to seem like the most attractive things in the world, simply because you're used to indulging in them. Still, you are more than capable of holding firm against their allure, especially if you focus on improving your self-discipline.

The simplest way to ensure you aren't tripped up by temptations is to avoid situations where these temptations might be present. As the old saying goes, "out of sight, out of mind." If you have biscuits, crisps, and other junk foods lying around the house, store them away in a pantry so you're not constantly looking at them. Most of the time, you will forget they're even there. Better yet, if no one else in your house is eating junk food, take some time to clear out your fridge and pantry and donate or throw away the offending items. If these foods aren't even in your home to begin with, you would have to go out to the store to get them, which serves as a barrier in between you and caving to temptations.

If you can't simply clear away all temptations from your life, you can employ other resistance strategies like distractions and focusing on your goals. Distractions only benefit you temporarily, but they can be just what you need to give yourself time to reconsider whether or not you really want to indulge. Do you really want to set back your progress for the sake of a double cheeseburger or some pie? Do you really want to put your health in jeopardy by skipping out on your workout routine? When you shift your focus to how failing to maintain your healthy habits will put your goals on the backburner, and what this means for your overall well-being, you will be far less likely to cave to temptations.

Create Goals and Rewards

Why do you want to live a healthier life? What level of physical fitness are you looking to achieve, and what would it enable you to do? Setting goals is an important part of any effort to become healthier, as they serve as indicators that you're making a real positive difference in your life. It can be difficult to stick to eating healthy if you don't really care one way or another, but if you're very motivated to prioritise your health because you have made it one of your goals, you're more likely to stick with healthy meals even when stress makes it difficult to ignore temptations.

When creating goals, make sure they are specific and measurable, ideally with a set time period. You can always adjust this length of time if complications arise, but try to pick something reasonable so you won't have to change it. Setting measurable goals means you must be able to know, at any time, how close you are to completing them. "Get healthy" isn't measurable because you haven't defined what 'healthy' means in this context. "Eat five servings of vegetables per day," "workout for 30 minutes each day," and "lose three stone in six months" all make your conditions for success clear, so you know how far you've progressed and how much further you have to go.

Don't forget to give your goals meaning by taking some time to reflect on why you want to achieve them. Maybe you want to lose weight and get in shape so you have more energy to go on walks with friends. You might want to eat healthier so you can chase away brain fog and do better at your job, allowing you to achieve greater success. Maintaining good health could mean the difference between still being able to play with your future grandkids and being held back by a chronic health condition. These rationales imbue your goals with far greater meaning and weight so you won't give up on them when times are tough. It isn't just about shedding weight or lowering your cholesterol. It's about creating and living the life that you want, now and in the future.

Finally, once you've set your goals and you start working towards them, create milestones for your progress and reward yourself when you reach them. If you want to lose three stone, celebrate when you lose one, and again when you're down two stone. If you're training to run a five-kilometre race, reward yourself whenever you improve upon your best time. Just make sure the reward is still healthy! You don't want to undo all the hard work you've done so far. Try positive reinforcement like spending a day with friends or sharing your accomplishments with others, or maybe spending some time curled up on the couch playing video games or reading your favourite book. These rewards shouldn't get in the way of your healthy habits, but it's always a good idea to celebrate your successes.

Seek Support

There will eventually come times when you run out of motivation and you need a little help sticking to your goals. This is perfectly normal. Just as you can get burnout at work or during a particularly hectic period of your life, you can also feel burnt out on your goals. When this happens, one of the best things you can do for yourself is asking for help, no matter how hard that might feel sometimes.

Self-reliance is important, but it shouldn't come at the cost of your mental or physical health. If you need some support from friends and family, that's perfectly fine. Be open and honest about your health goals, and ask your family to try to respect your decisions. If you're attending your mum's birthday party and they serve cake, explain that you're trying to eat healthier and you don't want to indulge right now, and that they shouldn't pressure you into eating it if you don't want to. If you want to eat healthier but you're not a great cook, ask the chefs in your life to teach you how to make a few basic, healthy meals. If you're struggling to stick to your workout routine, consider going on a run or hitting the gym with a friend by your side. When you have someone to hold you accountable and push you towards achieving your fitness goals, it is easier to stay on track and maintain your positive mindset.

Taking Control of Your Diet and Exercise

Far too often, diet and exercise levels seem to control us, not the other way around. We have every intention of working out that day, but an additional project at work leaves us too stressed to continue, and we simply shrug and decide we'll do it tomorrow. We go out to a restaurant with friends and we mean to order grilled chicken or fish, but we end up getting a burger or steak because of cravings. These are common fitness mistakes, and they all point to one central issue: our whims are in control of our actions, when we should be in control of our whims.

The process of taking back control involves reviewing your bad habits and their environmental triggers, just like you did in Chapter 6 for your maladaptive habits for mental health. As you begin to understand what kind of triggers prompt bad behaviours and make unhealthy cravings worse, you can become more adept at avoiding them or refocusing yourself on a new response to these cues, overriding the old bad habits.

Remember that this process should all come from a positive mental state that empowers you to make healthier choices. By using this positivity to your advantage, you can create new healthy habits that you *can* actually achieve, and you'll have far less trouble leaving the old ones behind. Everything begins with developing a better understanding of why you slip up and what you can do to better respond to these habit triggers in the future.

Understanding Your Bad Habit Triggers

If you're trying to lose weight and get in shape, or improve your eating habits to cut down on fat and cholesterol, there will likely be times where you're envious of all the progress someone else seems to be making while you're stagnating. Why does it seem easy for your friend to lose weight, while you feel like you're following all the same steps and seeing no results? How come your sister seems to have boundless energy that allows her to stay in the gym all day, but you can barely scrounge up the motivation to even think about getting on a treadmill?

Part of the reason for these differences might be that you are inadvertently allowing your bad habit triggers to flourish without your knowledge or consent. As previously mentioned, bad habits can be sneaky, and you might not even realise what parts of your environment make you more likely to reach for the biscuit jar or leave you feeling so mentally exhausted that physical exercise seems out of the question. Everyone has different struggles they encounter in life, but if you allow these hardships to take their toll on your physical well-being, you give them too much power over you. It is only by understanding what kinds of situations make you more likely to ignore your good intentions and return to your old bad habits that you can regain control and put a stop to unhealthy behaviours.

Make sure to pay attention to your stress and anxiety levels especially, as these are frequently connected to excessive or disordered eating, as well as a lack of motivation for exercise. When you feel low, you're more likely to revert back to the things that bring you comfort. If comfort comes in the form of watching trashy TV on the couch and eating a bag of crisps, you're strengthening the association between unhealthy behaviours and the pleasant feelings they provide. It may seem harmless, but you're actually convincing your brain that this is the only way you can let off steam, and that you couldn't possibly maintain the healthy habits you've established whenever you experience difficult moments in your life. This is a bad pattern you don't want to fall into, so it's important to interrupt it by finding a new way to react to stress and anxiety cues. For one, you could work to lower your anxious response to situations through positive thinking, meditation, and other mind-healthy habits. With less overall stress, you're less likely to indulge your bad habits. Alternatively, you can try turning a healthy habit into a stress reliever.

Healthy habits often feel like something you do when you're in a peaceful mental state, but this often isn't true. In fact, many people use exercise to blow off some steam when they're angry or upset. The feeling of moving your body gets your blood pumping the right way, and you feel more energised. You don't really want to hurt anyone, but using this excess energy to lift weights or jab at a punching bag can help you burn off some of the stress you're under. The same is true for cooking. The kitchen can be an especially frantic place, and having to keep track of all the steps for a complicated dish can really take your mind off whatever is bothering you. Cooking can help you alleviate your stress just the same as exercise, and at the end of it, you have a tasty, healthy meal. Try channelling any negative emotions into positive outcomes, and you'll find that it becomes much easier to stay on track with your diet and exercise habits even when things are going wrong in your life.

Creating a Positive Mental State

Where does the drive for change come from for you? Do you want to create a better life for yourself, or are you thinking about all the times someone had something nasty to say about your weight or physical ability? Are you consistently practicing self-love, or do you find yourself repeating many of those criticisms towards yourself? Some people believe they can bully themselves into getting healthy, but when change comes from negativity, it rarely lasts very long, and it never truly allows you to show yourself compassion and understanding when you need it most.

It is critical that your diet and exercise improvements come from a place of love, not one of self-hatred. You should feel very positive about your future goals, but you should also understand and accept where you are right now. It is okay to work to lose weight or to become more fit, but there's no reason that you have to hate your current body in order to achieve these goals. At the end of the day, you are your body, and showing yourself kindness is always a better practice than allowing yourself to be needlessly cruel.

Establishing a positive mental state is the best foundation you can give yourself for self-discipline and willpower. You want to change because you care about yourself, your health, and your overall well-being. This way of thinking will help you overcome just about any challenge. Not only do you have faith in yourself that you can stick to your healthy habits and achieve all the goals you've laid out for yourself, but you'll also have a source of positivity you can rely on whenever times get tough. The right mindset means everything, and nurturing optimism and kindness in your outlook on life will only help you accomplish your goals.

Taking Back the Reins

If you allow your current eating and exercise habits to continue unexamined, you might be giving them too much control over your health and your behaviours. If you want to eat healthy but you keep eating things you know you shouldn't, this means you are no longer the one in control of the situation, which can be a problem. It is incredibly difficult to start eating healthier and exercising more when you lack the control over your own life and actions to keep up these habits. If you don't take the necessary steps to make informed decisions about what you do and don't eat, how often you exercise, and other methods of taking care of your physical health like getting enough rest, your current habits are going to end up controlling you.

A positive mindset helps you find the inner strength to exert your authority and power over your bad habits and cravings. If you find that you're always looking at your habits in a negative light, you're quickly going to stop believing that you can make any changes at all. You'll resign yourself to making unhealthy choices consistently, assuming you're simply not strong enough to get in shape. However, this is completely untrue. Everyone is capable of adopting healthier habits and making the right choices for themselves, but to do so, you must believe you have the self-control to accomplish it. The more faith you have in yourself from your positive mindset, the easier it will be to make healthy choices and recognise the equally positive effect these choices are having on your body and mind. By consistently making the right choices, you'll be able to live a longer, happier life, and you'll have all the energy you need to pursue success in all forms.

Chapter 9: How to Be Successful

Success itself is sometimes a difficult concept to define, and it can vary between different people. Some might be perfectly content in a small house with a meagre income as long as they are surrounded by family, while others desire enough money that they can provide for themselves and their loved ones without having to ever worry about financial difficulties. Some find their true meaning and purpose in their work, while others look for different passions and hobbies that they want to excel in even if they're not being paid to do so. Some people prioritise getting in shape and carefully curating their appearance so they look a certain way, while others care little for how they look as long as they are healthy. These examples should make it abundantly clear that 'success' does not have a single correct definition. Instead, everyone has their own ideas about what will make them happy and what is worth pursuing for them. Therefore, to achieve success, you must decide exactly what this means for you.

What would make you happiest in life? What kind of things do you dream about, and what sort of people do you want to fill your life with? What is important to you, and what could you live without if it meant achieving other goals? Answering these questions requires some self-reflection, so you don't have to have a response right away. Still, it is a good idea to create your own personal definition of success so you can make sure you're always working towards it with every healthy choice you make.

One thing to keep in mind no matter where your pursuit of success leads you is that caring about your health is critical to achieving any goal you set out for yourself. Without the proper foundation of good mental health, you will struggle to stay motivated, to keep your stress levels under control, and to take advantage of new opportunities without fear and uncertainty getting in the way. Without taking care of yourself physically, you won't have the energy or focus you need to work on projects or to support your mental well-being. In this way, mental fitness is the basis upon which all forms of success must follow. To feel like you've achieved success with your life, you must have a passion and willingness to go after your dreams, a strategy for maintaining productivity, the right attitude, and of course, a goal to work towards.

Be Willing to Succeed

Consider this example: you and a co-worker are competing for the same promotion. You both make an effort to get noticed at work, but you doubt your own abilities, while your co-worker has no problem believing they will succeed. You want to speak up more often, but you're not sure if anyone wants to listen to what you have to say, so you end up staying quiet. You want to put yourself out there and try taking on more projects, but you assume you aren't capable of handling certain projects that are outside of your comfort zone, so you let these opportunities go. You worry that you'll be annoying your boss if you talk to them about the promotion, so you just hope they recognise your hard work and give it to you without the need for an awkward conversation. Meanwhile, your co-worker ensures their voice is heard at every meeting. They eagerly try out new things and give every project their all, even if it means doing something they've never done before. They have faith in themselves. They ask to meet with the boss and they explain how much the promotion would mean for them and why they would be a good fit, because they honestly believe they are more than capable of handling the job.

If you were your boss, who would you choose for the promotion? Nine times out of 10, the person who has confidence in their skills and their ability to succeed is going to actually follow through on that success, while the person who holds themself back with doubts isn't going to have nearly as happy of an outcome. You must ask yourself if you're really willing to succeed, and you must believe you are. Build up the confidence needed to try out anything you've never done before. Go into new experiences with the assumption that either they will pan out for you or, if something goes wrong, it will not be that big of a deal and you can recover. Stop allowing self-doubt and endless criticisms directed towards yourself to hold you back from doing a good job, or you're never going to give your pursuits your all. Remind yourself of how capable you are, and when you're in doubt, just think, "I've got this." You will be amazed at how much more you can accomplish when you simply start putting faith in yourself, and others will be much more receptive to your confidence.

View Yourself as Worthy of Success

One problem that holds many people back is that they don't see themselves as being worthy of success. For example, you might want your life to improve, but whenever an opportunity to apply for a better job arises, you decide you aren't ready for it without even bothering with the application process. You might think about the kinds of people who tend to find success and happiness in life and decide that's just not who you are. You might trick yourself into believing you're not deserving of happiness, positive relationships, or even love, especially self-love. Many of these dark thoughts can be symptomatic of underlying mental health conditions like depression, but some can arise in moments of weakness in otherwise healthy individuals. In either case, it's necessary to challenge these thoughts when they occur. Not only are they self-destructive, but they're also completely untrue.

When you allow these kinds of negative thoughts to take hold, you undermine your ability to recognise the good in yourself. Self-worth is a core skill, and yet it's something that isn't generally taught in schools. Therefore, you must learn to cultivate it on your own. This means making an effort to recognise when you're being unfair to yourself. Would you think so negatively if you were talking about one of your friends who was just trying to do their best? Would you be so cruel to your younger self if you could talk to them? If not, then why do you feel the need to hold yourself to harsher standards than you would anyone else? Remind yourself that you are worthy of love and respect at all times, no matter what you do, simply because you are human. Acknowledge that it is normal to make mistakes sometimes. Practice showing yourself gratitude and love for doing everyday things like taking care of yourself or for simply existing. You don't need to achieve success to see yourself as someone worth caring about, but you do need to see yourself as worthy of love in order to succeed.

Wiring Yourself for Winning

You might be surprised to hear that part of what might be holding you back from achieving success is that your brain just isn't wired for winning. The way you think about the idea of winning might very well be the thing that is complicating your ability to work towards victory. For example, you might tend to second-guess yourself, frequently seeking out the opinions of others and pausing before making decisions to scrutinise every angle of the issue. While it's important to put some consideration into big decisions, if you delay for too long, you might miss out on a big opportunity. Additionally, trying to predict all possible outcomes can lead you to think more negatively, as your brain starts picturing everything that could go wrong. Before long, you end up over-analysing your thoughts and actions, to the point that the version of yourself you show the rest of the world may not even be true to who you really are.

Over time, repeated negative thinking can train your brain to adopt a number of other harmful reasoning methods. For example, you might develop a tendency to overgeneralise, assuming that just because one thing goes wrong, the whole day is going to follow the same pattern. You might focus on completing a task in just one way, and if that method doesn't work out, you may stew in feelings of failure rather than trying to come up with a new way to handle the issue. You may tend to fixate on only the negative aspects of any situation rather than looking on the bright side. Every situation you encounter serves to train your brain to approach future situations in a similar way. If you consistently practice negative thinking, you're more likely to think negatively in the future, and you'll struggle to regain feelings of confidence and a willingness to try new things.

How do you break out of these habits? Just as you wired your brain to think negatively, you must rewrite it by wiring your brain to think positively. If you notice yourself getting carried away with a negative thought, pause and evaluate how accurate or fair that thought really is.

You might have started your day on the wrong foot by getting into a fight with your spouse, but does that mean you're going to have a bad day at work? Is it really accurate to say that your relationship with your spouse is ruined forever, or can you make an effort to apologise and find some common ground so you can remain positive during the rest of your day? If you pick up on your hesitancy to try something new, ask yourself why that is and what would realistically happen if you gave it your best shot. Are there any real negative consequences that are actually likely to happen, or are you psyching yourself out before you've even given yourself a chance?

When you pause to re-examine your negative thoughts, you can take a more rational approach to many situations. This will help you embrace the idea that achieving success is possible, especially if you remember to be patient with yourself when you need it and to only take on as much responsibility as you can handle.

Balancing Work and Home Life

Another aspect of a "poorly wired" brain that often trips people up is an inability to maintain a proper sense of balance in their lives. Specifically, work-life balance is often a point of contention. You might be very focused on achieving your professional goals, but if you're not careful, this can interfere with your ability to achieve your personal goals. Your work might run into the time you have to relax and unwind at home, and constant stress from deadlines can strain your relationships. As a result, you don't get the rest you need to operate at your most productive, and your ability to succeed both professionally and personally suffers for it.

Achieving and maintaining a good balance between work and relaxation is key for pursuing your goals while still respecting your mental health. Check in with yourself periodically, and if you're feeling stressed and overwhelmed, take a break and perform a calming activity.

Set hours when you work and hours when you're out of the office, and avoid letting these hours blend together. Take care of yourself and pay attention to your needs.

Stay properly hydrated. Remember to eat nutritious meals and squeeze in a bit of exercise when you can. By maintaining a sense of balance in your life, you can continue to work towards your goals without burning yourself out, and you give yourself a more robust line of defense against negative thoughts.

Adopt a Positive Mental State

Throughout this book so far, we have looked at ways of improving your mentality and adopting a more positive outlook. This has helped you to replace bad habits, alleviate anxiety, and improve your physical health. Now, these same ideas will help you take a better approach towards finding success in your life. As always, positive thinking will help give you the boost you need to achieve your goals if you know how to harness it properly.

Think back to the example from earlier in this chapter where you and a co-worker both wanted the same promotion. In the example, your boss paid more attention to your co-worker because they had the right mindset. They valued themselves and their work, which made others recognise the value in their efforts in return. They were also more willing to take risks and speak up, so they received more notice. This was only possible because of their positive mindset. To do the same in your own life, you must embrace positive thinking, as it will carry you towards success in every area of your life.

Improving Your Mindset

The idea of improving your mindset can often be complicated by the many barriers that stand between you and pursuing proper treatment for any mental health issues you may be experiencing. It is hard to believe in yourself when your depressive thoughts undermine your confidence. It is difficult to put yourself out there when your anxiety has convinced you that no one wants to hear you speak.

Unfortunately, social stigmas can perpetuate negative mindsets so you are unable to achieve the kind of support you need to really embrace positivity. It's hard to talk about any mental health issues you might be experiencing, let alone to actually ask for help with making improvements to your mindset. It is necessary to first and foremost recognise that there is nothing wrong with you needing a little extra assistance. Mental health problems happen, and while it is important to work to fix them, hiding them out of shame and fear does nothing to help you develop greater mental strength. When you let go of negative thoughts about your mental state, you'll often find it infinitely easier to make improvements to your mindset.

Improving your mindset means learning to recognise that everyone has flaws, even you, and that's no reason to be unfair to yourself. It means looking for your positive qualities and harnessing this information to approach problems from your own unique angle to achieve success. It means accepting that you will have good days and bad days, and that even the bad days can be productive ones if you are kind to yourself and do what you can. This is a gradual process, but as you employ the tools and strategies you've learned so far to alleviate the stress and tension that contribute to negative thoughts, it should become easier to truly embrace positivity. Once you have established a positive outlook, you can then use this to support yourself in your endeavours, especially where it helps you show yourself kindness and compassion.

Being Compassionate Towards Yourself

Few people have ever accomplished anything of value by constantly berating themselves. Think of how you feel when your boss notices a mistake you have made and decides to take out their frustration on you. When someone yells at you, you feel hurt and often demoralised. Rather than wanting to work harder to prove yourself next time, you might want to quit, throwing your hands up and deciding that you're just not cut out for this kind of work after all. Excessive criticisms and yelling only breeds negativity, which can become a dream-killer. Now, consider how you feel when your boss notices you struggling and, instead of attacking you for it, tries to find out what went wrong and offer their support. In this case, you might be more willing to try again, with the benefits of some assistance and your improved self-confidence. You recognise that your mistake, while unfortunate, wasn't the end of the world, and you're more motivated to correct it rather than defaulting to negativity. In short, when others show you compassion, you perform better.

When a workplace culture breeds negativity, it is no surprise when you begin to internalise that negativity as well.

You are the boss of your brain. If you yell at yourself when you get things wrong, you're going to demoralise yourself pretty quickly, and you'll likely give up on your dreams. Success will be just out of your grasp because you told yourself that you were a good-for-nothing and you were never going to become anything in life. If you instead provide yourself with the support and understanding you need, you will flourish. You'll tackle every challenge head-on without fear, because you know that failure isn't permanent unless you allow it to be. You understand that success is always within your grasp as long as you keep trying, and you see the value in yourself even if you fall a little short of your goals. By practicing compassion, you support yourself, and you support your ability to chase your dreams.

Pursue Success Relentlessly

Once you have a good idea of what you want to accomplish, it's time to go out and get it. There will be moments when you feel worn down or discouraged, often because you are working very hard but not seeing results right away. Maybe you want to feel more comfortable in social settings, but even though you have been implementing some helpful strategies like breathing exercises and getting better rest to improve your emotional control, you feel like you still have a long way to go. Typically, this kind of progress isn't instantaneous. In fact, just about anything that is worth doing will take a great deal of time and effort to accomplish, so it's normal not to see results right away. This is only a problem if you allow this lack of results to discourage you.

Maintaining your motivation can be difficult, but it becomes a little easier when you keep track of how far you have already come. You might not be able to completely dispel your anxiety from a few weeks of trying to overcome it, but chances are that all the time you spend acclimating yourself to the kinds of situations that would otherwise spike your anxiety will still make a notable difference. You may still feel on-edge, but you might not be quite so tempted to completely flee the scene. You might still have anxious thoughts, but you know how to quell them. With each attempt you make, you get a little bit better, and a little bit closer to your goals. Recognise everything you've done so far and allow yourself to feel proud of all you've accomplished, even if you still have a way to go before reaching your overarching goal.

Remember that, as previously mentioned, slipping up and backsliding happens from time to time. It's nothing to be ashamed of. If you show yourself kindness when you make a mistake, you'll have an easier time recovering and trying again, and that's what really matters. You can fail a thousand times, but if you pick yourself up a thousand times, then none of those failures are permanent. Who knows? Since you're better equipped with the knowledge that each of those failures has given you, you might just succeed on the next attempt.

Live for Yourself, Not for Others

Every year, thousands of people go to university to learn how to become doctors and lawyers. Every year, a large percentage of those kids end up dropping out and pursuing a different career path. Why is this? Is it really just that these degrees are difficult, or is there something more to it?

One significant contributing factor could be that many of these students don't actually feel passionate about becoming a doctor or a lawyer. They might only pursue these jobs because they want to make money, or because their parents want them to have a respectable job. However, when you're working to achieve what someone else wants, not what you want, you're not likely to remain motivated for very long. When things get difficult, your willpower will wane, and you'll find that you are no longer willing to put yourself through so much turmoil for the sake of achieving something you never wanted in the first place.

You need to be passionate about your definition of success, and you must achieve it for your own sake, not for the sake of others. This is another example of finding value in yourself and your own wishes. If you take all your cues from other people, you won't be passionate about what you're doing, and you're more likely to fail. Even if you do succeed, your victory will feel hollow, as it's not something you desire. If you want to achieve real success in your life, you need to live for yourself. When you are pursuing dreams that you are genuinely invested in, all the hard work will feel easy. To remain motivated, you'll only need to remind yourself of what waits for you on the other side and the kind of life you'll have when you achieve your goals. Through finding success, you will also start to find the things in life that will make you happy.

Chapter 10: How to Be Happy

Poor mental fitness can make it incredibly difficult to find a sense of happiness and fulfilment in your life. This isn't just limited to people experiencing conditions like depression, who will often feel less excitement and passion for things that once gave them joy, but depressive feelings can be part of the problem. If you are constantly suffering from emotions that are running out of control, leaving you stressed and unable to function normally, you're going to find it very difficult to calm down and experience the moment. You may feel lost, confused, and upset on a regular basis. In short, the burden of emotional stress can prevent you from really feeling happy.

While this is certainly a cause for concern, it doesn't have to be a permanent problem. As you learn to improve your mental health, strengthening your emotional control and letting go of some of the stress that plagues you, you will start to find enjoyment in life again. You'll feel more passionate about fun activities, and you'll likely be more open to spending time with friends and family just for the sake of enjoying yourself. Your daily activities often feel more personally fulfilling, even if you don't necessarily love your job, and you learn to find pride in what you do rather than always putting yourself down. These are just a few of the many benefits of strengthening your mind, but they represent the idea that better mental fitness leads to greater happiness.

The path to happiness and fulfilment isn't always a straightforward one, but when you work to embrace healthier habits and try to see the world from a more positive point of view, you are sure to make progress down this road. Happiness involves many different components, each of which must be addressed to really feel at peace with yourself. You must learn to be self-confident and to feel secure in the knowledge of your own capabilities. You must let negativity go so it cannot burden you any longer. You must stop deferring to the wants and needs of others, and instead start listening to your own thoughts and feelings. Finally, you must learn to take charge of your own happiness, as this is the best way for you to truly improve and grow.

Self-Confidence and Security

Self-confidence and a related concept, self-esteem, are absolutely critical to your happiness. If you cannot express feelings of love and respect for yourself and your efforts in life, you will have a hard time truly being happy about any of your accomplishments. At the same time, if you tie your self-worth entirely to something like your performance at work or another skill, you may struggle to maintain your confidence when you make mistakes. It's important to learn how to recognise and appreciate your talents, as well as working to improve yourself each day, while still accepting yourself for who you are.

There are plenty of ways for you to reinforce your self-confidence every day. Start by acknowledging your efforts and allowing yourself to feel proud of all the good work you do. This can be anything from performing well at work to more social pursuits like helping a friend feel better when they need someone to talk to. Pay attention to where your natural talents lie, and acknowledge that there are things you're good at, even when you're feeling down in the dumps. Maintain a growth mindset and remember that if there is anything you're not so good at, you can work to change that. You are capable of much more than you know, and you'll only learn this lesson for yourself when you make a point to improve your self-esteem.

One of the most important things to avoid when building up your self-confidence is comparing yourself to others. This often comes from a deep-seated sense of insecurity. You want to feel like you're better than others rather than accepting that everyone has their strengths and weaknesses. If you focus all your mental energy on keeping track of what other people are doing and seeing whether or not you measure up, you'll never be able to recognise the value of your own efforts. Even if you aren't as good at a certain skill as someone else, or if you don't seem to have the dream life they have, this isn't a reason to ignore your talents and the good parts of your life. Practice being grateful for what you have, even if it is not as much as what others have. Be proud of what you can do, even if you're still learning. When you stop trying to live up to the standards of others, you can start really appreciating yourself, and you'll let go of the insecurities that undermine your sense of confidence.

Finding Relief From Negativity

Having a negative outlook in life will only get you so far. When you are constantly looking for something to criticise or complain about, you pass up a number of opportunities to recognise the sources of positivity in your life. This, in turn, can put a damper on your mood, as you are frequently grumpy, stressed, and upset. If you ignore all the wonderful things in your life, it will be no big surprise when you struggle to experience happiness on a day-to-day basis.

Making an effort to practice gratitude every day can help you overcome a tendency towards negative thinking. Being grateful means recognising all the good things in your life, as well as accepting that while bad things may happen, these events are never permanent, and they don't have to take precedent over the more positive events. Even if you are experiencing a very difficult moment of your life, there are still plenty of things to be grateful for. As an example, let's say you just lost your job and you're having trouble getting back on your feet. If you focus on the pain of losing your job and how angry and upset you are, you'll only succeed in making yourself miserable. You'll hardly improve your chances of getting a new job, as you'll likely have thoughts such as, "I'll just get fired from this one too," or "Why should I even bother applying when I'm such a failure?" These are hardly motivating thoughts, and they can seriously interfere with your ability to recover after difficult life events.

Now, consider how the same situation might go if you let go of the negativity and instead focused on being grateful for what you still have. Maybe you made a point to start saving while you were at your previous job, so you're thankful you have enough money to last you a few months. Maybe you're grateful that your friends and family are offering you their condolences and support. You might also be grateful for everything you learned at your previous job. While it may have ended poorly, you still gained experience, you might have made some friends, and you may have learned something about the kind of company you want to avoid in the future if you had a bad experience. Even appreciating that you still have a roof over your head and warm food to eat can help you let go of some of your hurt feelings. This allows you to come to terms with what happened and move forward with your life, seeking greener pastures.

Gratitude alone can't change your situation, but it can help you bounce back and leave you with a more positive disposition. Learning how to let go of negative feelings can make all the difference in your ability to find happiness in your life, no matter what you experience.

Learn to Put Yourself First

If you experience anxious feelings or low self-esteem, you may have a tendency to put the needs of others above your own. If you and your spouse have a difference of opinion, you might decide to defer to their point of view because you assume they know better than you. If your parents always wanted you to pursue a certain career, you might do so out of a sense of obligation to them for raising you. You may even find yourself taking a more submissive role in the workplace or amongst friends. While it is good to consider the opinions of others and how certain actions might hurt or help them, it's not such a good idea to forego your own sense of happiness in favour of making others happy. Making compromises is fine, but when these compromises require you to sacrifice your time, energy, physical well-being, or mental health, you're giving up much more than you gain, and you're not being fair to yourself.

As mentioned in the previous chapter, to feel happy about the direction your life is going, you must make choices for your own sake. You need to live for yourself and value your own opinions and feelings. It isn't healthy to let others walk over you, and this isn't a fulfilling way to live either. If you never pursue the things you're passionate about, you won't find the work you're doing important, and you'll likely struggle to achieve true happiness.

It can be very difficult to start putting yourself first if you're used to letting other people take charge of your life. This is especially true if you were ever in an abusive relationship where someone tried to exert complete control over your actions. You may be uncertain about whether or not you'll make the right choices, and you'll likely have trouble being assertive, particularly when you disagree with others. However, with some practice, you can begin finding value in yourself and learning to recognise your wants and needs. Start with small decisions, like choosing your outfit based on what you personally would like to wear, without worrying about how others will perceive that choice. Get into the habit of making choices whenever possible, even if you are just picking between two options. Through this process, you will learn to trust your own intuition, and you'll get used to the feeling of having control over the direction of your life. From here, you can move on to more significant decisions like career moves and ways of life that will make you happier. By putting yourself first, you'll be able to lead a more personally fulfiling life.

Self-Serving, Not Selfish

It's easy to make the mistake of conflating the idea of putting yourself first with the notion of acting selfishly and maliciously. You might feel like it's unfair to prioritise your own opinion, which can keep you from really embracing this new outlook on life. However, this concept comes from a misunderstanding. You don't really have to be selfish to put yourself first. You just need to be self-serving in a positive manner instead.

It's not necessarily a bad thing to be a little self-centred. In fact, it can be much more harmful to ignore your thoughts and feelings, as you have learned. When you start paying more attention to your wants and needs, you'll develop a greater sense of agency over your life. Being self-centred means centering your life on yourself, not acting to the exclusion of others but instead acting in a way that brings you joy and a sense of accomplishment. If you centre the thoughts and experiences of someone else, spending your whole life trying to please them, you fail paying attention to what you need. By refocusing on your own opinions, feelings, and values, you can live in a way that is true to yourself, and experience greater happiness as a result.

Take Responsibility for Your Happiness

When you learn to give yourself greater agency over your own life, you also start to take responsibility for your happiness. Ultimately, no one else is in charge of making you happy. If you are miserable in your current job, your boss may be the cause of your woes, but it's not necessarily their job to fix it. Only you can make the appropriate changes to make yourself happy again, whether this means finding a new way to work, adopting a better mindset, or simply switching to a more fulfilling job.

If you spend your life waiting around for someone to sweep away all your troubles, you're going to be waiting for a very long time. You are not Cinderella. You cannot wait for a prince to come and take you away from your cruel step-mother and step-sisters so you can enjoy a life of luxury. If you want your life to change for the better, that means taking action yourself. Becoming mentally fit is the first piece of the puzzle of improving your life and feeling happy with where you are. Only you can change your life and embrace the things that make living worthwhile.

Creating the Life You Want to Live

Creating the life you desire most for yourself means defining success and working towards it. It means getting a handle on your emotions, paying attention to your feelings without being completely overwhelmed by them. It means learning to try new things and managing the associated stress. Living your dream life also includes adopting healthy habits for both your mind and your body, as well as relying on your positive mindset to get you through the difficult times. Through these actions, you can begin to forge a life that allows you to be both healthy and happy.

This process will not always be easy. There will be many times when you may want to throw in the towel and attribute your lack of success to a cruel, unfair world. You'll have moments when you're upset and frustrated alongside the moments of calm and peacefulness. You won't experience immediate results. However, by making continual positive changes one step at a time, you'll gradually start to build the kind of life you want to live. You'll experience self-fulfilment, and you'll start to feel excited about the future again. Perhaps best of all, as you continue striving to be the best version of yourself at all times, you'll be happy. This is the true benefit of mental fitness. By striving for your ideal life and learning to love yourself along the way there, you'll see that happiness may not be such a far-off dream after all. Instead, you'll find that it's entirely within your reach.

Conclusion

At this point, you have learned all the skills and strategies you need to completely turn your mental health around. You've performed an honest analysis of your mental well-being, identified any problems that are holding you back from living your best life, and started thinking about how you can overcome these problems. You've discovered the power of good habits for your mind as well as your body, and you've learned useful methods for alleviating stress and anxiety whenever they may arise. By creating a positive mindset, you can pursue success and happiness—it all starts with taking action.

I know you are more than capable of creating this kind of positive change in your life. I am positive that if you follow the advice outlined in this book and make an honest effort to improve your mental fitness, you will succeed, and you will discover you are so much stronger than you once thought. Now, you need to believe the same about yourself. When you have faith in your own abilities, you will see just how much you can accomplish as you put these strategies into practice.

I appreciate the time you have spent reading this book, and I hope that the advice contained within has helped you create a plan for improving your mental strength.

If you want to learn more about the topics we've covered, visit https://thehealthandfitnesscoach.co.uk or https://thehealthandfitnesscoach.com for more information on bespoke mental health action plans. There, you will also find many resources that will help you on your diet and physical health journey to complement the work you're doing on your mental health.

I wish you the best of luck in your pursuits to strengthen your mind and embrace a positive mindset. I know that with hard work, you will create the kind of lifestyle you want for yourself, and you can leave the burdens of negativity, stress, and fear behind. If you follow the principles outlined within this book and take action, you will feel empowered to make lasting positive changes in your life, and you will experience success, happiness, and true freedom from worry.

References

Anxiety and Depression Association of America. (n.d.). *Physical activity reduces stress.* https://adaa.org/understanding-anxiety/related-illnesses/other-related-conditions/stress/physical-activity-reduces-st

---

Becker, S. J., Swenson, R., Esposito-Smythers, C., Cataldo, A., & Spirito, A. (2015, Dec. 1). Barriers to seeking mental health services among adolescents in military families. *Professional Psychology: Research and Practice, 46*(6), pp. 504-513. https://doi.org/10.1037/a0036120

Dweck, C. S. (2006). *Mindset: The new psychology of success*. Random House.

Elsesser, K. (2020, Oct. 2). *The debate on power posing continues: Here's where we stand.* Forbes. https://www.forbes.com/sites/kimelsesser/2020/10/02/the-debate-on-power-posing-continues-heres-where-we-stand/?sh=68107d3202ee

Goleman, D. (1995). *Emotional intelligence: Why it can matter more than IQ.* Bloomsbury Publishing.

Harvard Men's Health Watch. (2021, Mar. 30). *Sour mood getting you down? Get back to nature.* Harvard Health Publishing. https://www.health.harvard.edu/mind-and-mood/sour-mood-getting-you-down-get-back-to-nature

---

Mental Health Foundation. (2015, Nov.). *Stigma and discrimination.* https://www.mentalhealth.org.uk/a-to-z/s/stigma-and-discrimination

---

Saxby, D. E. & Repetti, R. (2010, Jan.). No place like home: Home tours correlate with daily patterns of mood and cortisol. *Personality & Social Psychology Bulletin, 36*(1), 71-81.

SleepScore Labs. (2020, Nov. 15). *How to get more deep sleep: 5 tips and tricks.* https://www.sleepscore.com/extend-deep-sleep/

Society for Personality and Social Psychology. (2014, Aug. 8). *How we form habits, change existing ones.* Science Daily. https://www.sciencedaily.com/releases/2014/08/140808111931.htm

---

Truong, T. (2019, Jan. 14). *How clutter affects your health.* ABC News. https://abcnews.go.com/Health/clutter-affects-health/story?id=60367240

Wein, H. & Hicklin, T. (2012, Jan.). *Breaking bad habits: Why it's so hard to change.* National Institutes of Health. https://newsinhealth.nih.gov/2012/01/breaking-bad-habits

Printed in Great Britain
by Amazon